Water Resources Engineering: Planning and Management

Water Resources Engineering: Planning and Management

Edited by Ellie Legrand

SYRAWOOD
PUBLISHING HOUSE

New York

Published by Syrawood Publishing House,
750 Third Avenue, 9th Floor,
New York, NY 10017, USA
www.syrawoodpublishinghouse.com

Water Resources Engineering: Planning and Management
Edited by Ellie Legrand

International Standard Book Number: 978-1-68286-619-1 (Hardback)

Cataloging-in-Publication Data

Water resources engineering : planning and management / edited by Ellie Legrand .
    p. cm.
Includes bibliographical references and index.
ISBN 978-1-68286-619-1
1. Hydraulic engineering. 2. Hydrology. 3. Water resources development.
4. Waterworks--Planning. 5. Waterworks--Management. I. Legrand, Ellie.
TC409 .W38 2018
627--dc23

TABLE OF CONTENTS

PREFACE

Water is the most abundant resource on our planet, but fresh water is scarce. The need of fresh, useful and drinkable water is increasing rapidly but the resources are not. Thus, water resource planning and management is an important process. It refers to the proper management, planning, distribution and treatment of water so that it is used in an optimum way. It is a pivotal part of water-cycle management. The aim of this book is to provide in-depth knowledge about the processes of water resource management and planning. It presents topics, which are of utmost importance and bound to provide incredible insights to readers.

To facilitate a deeper understanding of the contents of this book a short introduction of every chapter is written below:

Chapter 1- Water, like energy, is neither created nor destroyed. It changes its phase as is evident from the water cycle. The distribution of water is uneven and the majority of it is found in ocean bodies. However, oceans are saline and fresh water amounts to just a fraction of water available on Earth. This is an introductory chapter which will introduce briefly all the significant aspects of the sources of water.

Chapter 2- Water resources modeling is extremely significant and useful when preservation of water and its quality is concerned. A few models such as river basin planning and management, water distribution system, water quality modeling, etc. are discussed in this chapter. In river basin, surface water and groundwater are submerged together, due to which its planning and management gains importance. The major components of modeling in water resources are discussed in this section.

Chapter 3- Irrigation adversely impacts soil and water and contributes to their pollution. Its direct effects include increased evaporation, decline in river flow, and also affect moisture, and other environmental factors of the area. Indirect effects of irrigation are soil salination, waterlogging, saltwater intrusion etc. The chapter serves as a source to understand the major categories related to impacts of irrigation.

Chapter 4- The presence of substances in water that is harmful to biotic and abiotic elements is called water pollution. Water pollution can be classified into surface water pollution and groundwater pollution. While the former concentrates on water pollution at the surface level, the latter studies the pollution keeping groundwater in focus. The chapter closely examines the principle sources of surface water pollution to provide an extensive understanding of the subject.

I owe the completion of this book to the never-ending support of my family, who supported me throughout the project.

Editor

Water Resources: An Overview

Water, like energy, is neither created nor destroyed. It changes its phase as is evident from the water cycle. The distribution of water is uneven and the majority of it is found in ocean bodies. However, oceans are saline and fresh water amounts to just a fraction of water available on Earth. This is an introductory chapter which will introduce briefly all the significant aspects of the sources of water.

Hydrological Cycle

Water in our planet is available in the atmosphere, the oceans, on land and within the soil and fractured rock of the earth's crust Water molecules from one location to another are driven by the solar energy. Moisture circulates from the earth into the atmosphere through evaporation and then back into the earth as precipitation. In going through this process, called the Hydrologic Cycle, water is conserved – that is, it is neither created nor destroyed.

Hydrologic cycle

It would perhaps be interesting to note that the knowledge of the hydrologic cycle was known at least by about 1000 BC by the people of the Indian Subcontinent. This is reflected by the fact that one verse of Chhandogya Upanishad (the Philosophical reflections of the Vedas) points to the following:

"The rivers… all discharge their waters into the sea. They lead from sea to sea, the clouds raise them to the sky as vapour and release them in the form of rain…"

The earth's total water content in the hydrologic cycle is not equally distributed.

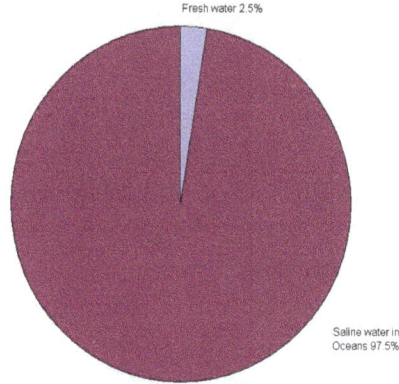

Fresh water 2.5%

Saline water in Oceans 97.5%

Total global water content

The oceans are the largest reservoirs of water, but since it is saline it is not readily usable for requirements of human survival. The freshwater content is just a fraction of the total water available.

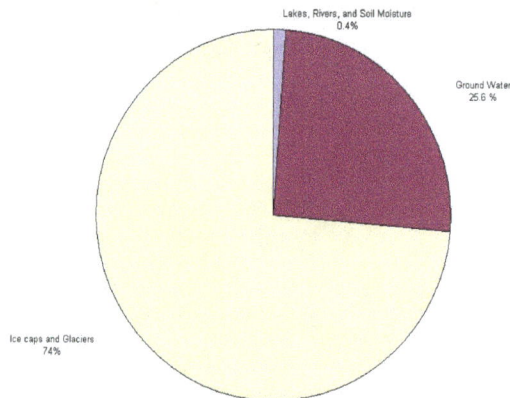

Lakes, Rivers, and Soil Moisture 0.4%

Ground Water 25.6 %

Ice caps and Glaciers 74%

Global fresh water distribution

Again, the fresh water distribution is highly uneven, with most of the water locked in frozen polar ice caps.

The hydrologic cycle consists of four key components

1. Precipitation

2. Runoff

3. Storage

4. Evapotranspiration

Precipitation

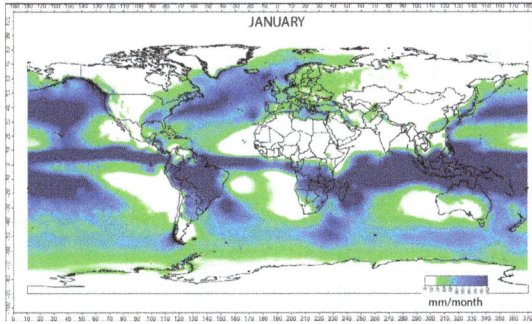

Long-term mean precipitation by month

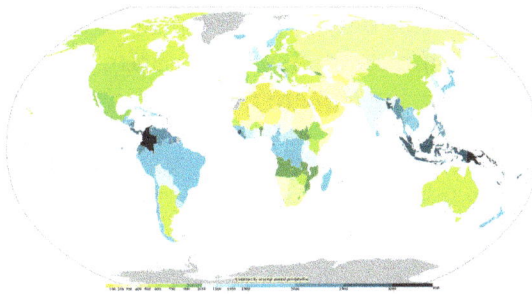

Countries by average annual precipitation

In meteorology, precipitation is any product of the condensation of atmospheric water vapor that falls under gravity. The main forms of precipitation include drizzle, rain, sleet, snow, graupel and hail. Precipitation occurs when a portion of the atmosphere becomes saturated with water vapor, so that the water condenses and "precipitates". Thus, fog and mist are not precipitation but suspensions, because the water vapor does not condense sufficiently to precipitate. Two processes, possibly acting together, can lead to air becoming saturated: cooling the air or adding water vapor to the air. Precipitation forms as smaller droplets coalesce via collision with other rain drops or ice crystals within a cloud. Short, intense periods of rain in scattered locations are called "showers."

Moisture that is lifted or otherwise forced to rise over a layer of sub-freezing air at the surface may be condensed into clouds and rain. This process is typically active when freezing rain is occurring. A stationary front is often present near the area of freezing rain and serves as the foci for forcing and rising air. Provided necessary and sufficient atmospheric moisture content, the moisture within the rising air will condense into clouds, namely stratus and cumulonimbus. Eventually, the cloud droplets will grow large enough to form raindrops and descend toward the Earth where they will freeze on contact with exposed objects. Where relatively warm water bodies are present, for example due to water evaporation from lakes, lake-effect snowfall becomes a concern downwind of the warm lakes within the cold cyclonic flow around the backside of extratropical cyclones. Lake-effect snowfall can be locally heavy. Thundersnow is possible within a cyclone's comma head and within lake effect precipitation bands. In mountainous areas, heavy precipitation is possible where upslope flow is maximized within windward sides of the terrain at elevation. On the leeward side of mountains, desert climates can exist due to the dry air caused by compressional heating. The movement of the monsoon trough, or intertropical convergence zone, brings rainy seasons to savannah climes.

Precipitation is a major component of the water cycle, and is responsible for depositing the fresh water on the planet. Approximately 505,000 cubic kilometres (121,000 cu mi) of water falls as precipitation each year; 398,000 cubic kilometres (95,000 cu mi) of it over the oceans and 107,000 cubic kilometres (26,000 cu mi) over land. Given the Earth's surface area, that means the globally averaged annual precipitation is 990 millimetres (39 in), but over land it is only 715 millimetres (28.1 in). Climate classification systems such as the Köppen climate classification system use average annual rainfall to help differentiate between differing climate regimes.

Precipitation may occur on other celestial bodies, e.g. when it gets cold, Mars has precipitation which most likely takes the form of frost, rather than rain or snow.

Types

A thunderstorm with heavy precipitation

Precipitation is a major component of the water cycle, and is responsible for depositing most of the fresh water on the planet. Approximately 505,000 km (121,000 mi) of water falls as precipitation each year, 398,000 km (95,000 cu mi) of it over the oceans. Given the Earth's surface area, that means the globally averaged annual precipitation is 990 millimetres (39 in).

Mechanisms of producing precipitation include convective, stratiform, and orographic rainfall. Convective processes involve strong vertical motions that can cause the overturning of the atmosphere in that location within an hour and cause heavy precipitation, while stratiform processes involve weaker upward motions and less intense precipitation. Precipitation can be divided into three categories, based on whether it falls as liquid water, liquid water that freezes on contact with the surface, or ice. Mixtures of different types of precipitation, including types in different categories, can fall simultaneously. Liquid forms of precipitation include rain and drizzle. Rain or drizzle that freezes on contact within a subfreezing air mass is called "freezing rain" or "freezing drizzle". Frozen forms of precipitation include snow, ice needles, ice pellets, hail, and graupel.

How Air becomes Saturated

Cooling Air to its Dew Point

Late-summer rainstorm in Denmark

Lenticular cloud forming due to mountains over Wyoming

The dew point is the temperature to which a parcel must be cooled in order to become saturated, and (unless super-saturation occurs) condenses to water. Water vapour normally begins to condense on condensation nuclei such as dust, ice, and salt in order to form clouds. An elevated portion of a frontal zone forces broad areas of lift, which form clouds decks such as altostratus or cirrostratus. Stratus is a stable cloud deck which tends to form when a cool, stable air mass is trapped underneath a warm air mass. It can also form due to the lifting of advection fog during breezy conditions.

There are four main mechanisms for cooling the air to its dew point: adiabatic cooling, conductive cooling, radiational cooling, and evaporative cooling. Adiabatic cooling occurs when air rises and expands. The air can rise due to convection, large-scale atmospheric motions, or a physical barrier such as a mountain (orographic lift). Conductive cooling occurs when the air comes into contact with a colder surface, usually by being blown from one surface to another, for example from a liquid water surface to colder land. Radiational cooling occurs due to the emission of infrared radiation, either by the air or by the surface underneath. Evaporative cooling occurs when moisture is added to the air through evaporation, which forces the air temperature to cool to its wet-bulb temperature, or until it reaches saturation.

Adding Moisture to the Air

The main ways water vapour is added to the air are: wind convergence into areas of upward motion, precipitation or virga falling from above, daytime heating evaporating water from the surface of oceans, water bodies or wet land, transpiration from plants, cool or dry air moving over warmer water, and lifting air over mountains.

Formation

Raindrops

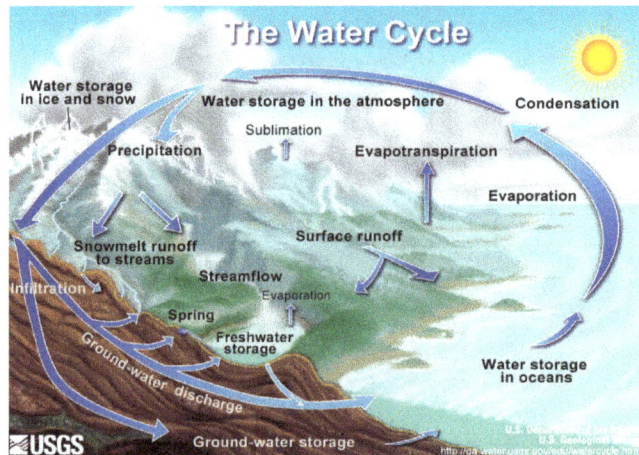
Condensation and coalescence are important parts of the water cycle.

Coalescence occurs when water droplets fuse to create larger water droplets, or when water droplets freeze onto an ice crystal, which is known as the Bergeron process. The fall rate of very small droplets is negligible, hence clouds do not fall out of the sky; precipitation will only occur when these coalesce into larger drops. When air turbulence occurs, water droplets collide, producing larger droplets. As these larger water droplets descend, coalescence continues, so that drops become heavy enough to overcome air resistance and fall as rain.

Raindrops have sizes ranging from 0.1 millimetres (0.0039 in) to 9 millimetres (0.35 in) mean diameter, above which they tend to break up. Smaller drops are called cloud droplets, and their shape is spherical. As a raindrop increases in size, its shape becomes more oblate, with its largest cross-section facing the oncoming airflow. Contrary to the cartoon pictures of raindrops, their shape does not resemble a teardrop. Intensity and duration of rainfall are usually inversely related, i.e., high intensity storms are likely to be of short duration and low intensity storms can have a long duration. Rain drops associated with melting hail tend to be larger than other rain drops. The METAR code for rain is RA, while the coding for rain showers is SHRA.

Ice Pellets

An accumulation of ice pellets

Ice pellets or sleet are a form of precipitation consisting of small, translucent balls of ice. Ice pellets are usually (but not always) smaller than hailstones. They often bounce when they hit the ground, and generally do not freeze into a solid mass unless mixed with freezing rain. The METAR code for ice pellets is PL.

Ice pellets form when a layer of above-freezing air exists with sub-freezing air both above and below. This causes the partial or complete melting of any snowflakes falling through the warm layer. As they fall back into the sub-freezing layer closer to the surface, they re-freeze into ice pellets. However, if the sub-freezing layer beneath the warm layer is too small, the precipitation will not have time to re-freeze, and freezing rain will be the result at the surface. A temperature profile showing a warm layer above the ground is most likely to be found in advance of a warm front during the cold season, but can occasionally be found behind a passing cold front.

Hail

A large hailstone, about 6 centimetres (2.4 in) in diameter

Like other precipitation, hail forms in storm clouds when supercooled water droplets freeze on contact with condensation nuclei, such as dust or dirt. The storm's updraft blows the hailstones to the upper part of the cloud. The updraft dissipates and the hailstones fall down, back into the updraft, and are lifted again. Hail has a diameter of 5 millimetres (0.20 in) or more. Within METAR code, GR is used to indicate larger hail, of a diameter of at least 6.4 millimetres (0.25 in). GR is derived from the French word grêle. Smaller-sized hail, as well as snow pellets, use the coding of GS, which is short for the French word grésil. Stones just larger than golf ball-sized are one of the most frequently reported hail sizes. Hailstones can grow to 15 centimetres (6 in) and weigh more than 500 grams (1 lb). In large hailstones, latent heat released by further freezing may melt the outer shell of the hailstone. The hailstone then may undergo 'wet growth', where the liquid outer shell collects other smaller hailstones. The hailstone gains an ice layer and grows increasingly larger with each ascent. Once a hailstone becomes too heavy to be supported by the storm's updraft, it falls from the cloud.

Snowflakes

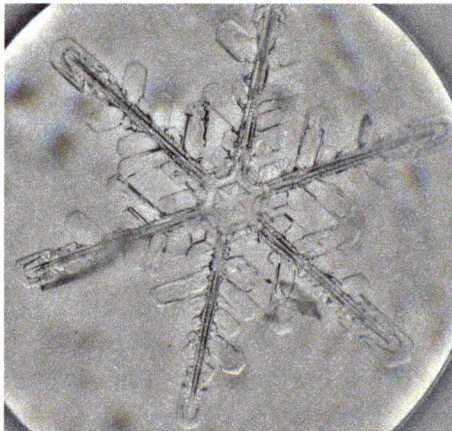

Snowflake viewed in an optical microscope

Snow crystals form when tiny supercooled cloud droplets (about 10 μm in diameter) freeze. Once a droplet has frozen, it grows in the supersaturated environment. Because water droplets are more numerous than the ice crystals the crystals are able to grow to hundreds of micrometers in size at the expense of the water droplets. This process is known as the Wegener–Bergeron–Findeisen process. The corresponding depletion of water vapour causes the droplets to evaporate, meaning that the ice crystals grow at the droplets' expense. These large crystals are an efficient source of precipitation, since they fall through the atmosphere due to their mass, and may collide and stick together in clusters, or aggregates. These aggregates are snowflakes, and are usually the type of ice particle that falls to the ground. Guinness World Records list the world's largest snowflakes as those of January 1887 at Fort Keogh, Montana; allegedly one measured 38 cm (15 inches) wide. The exact details of the sticking mechanism remain a subject of research.

Although the ice is clear, scattering of light by the crystal facets and hollows/imperfections mean that the crystals often appear white in color due to diffuse reflection of the whole spectrum of light by the small ice particles. The shape of the snowflake is determined broadly by the temperature and humidity at which it is formed. Rarely, at a temperature of around −2 °C (28 °F), snowflakes can form in threefold symmetry—triangular snowflakes. The most common snow particles are visibly irregular, although near-perfect snowflakes may be more common in pictures because they are more visually appealing. No two snowflakes are alike, which grow at different rates and in different patterns depending on the changing temperature and humidity within the atmosphere that the snowflake falls through on its way to the ground. The METAR code for snow is SN, while snow showers are coded SHSN.

Diamond Dust

Diamond dust, also known as ice needles or ice crystals, forms at temperatures approaching −40 °C (−40 °F) due to air with slightly higher moisture from aloft mixing with colder, surface based air. They are made of simple ice crystals that are hexagonal in shape. The METAR identifier for diamond dust within international hourly weather reports is IC.

Causes

Frontal Activity

Stratiform or dynamic precipitation occurs as a consequence of slow ascent of air in synoptic systems (on the order of cm/s), such as over surface cold fronts, and over and ahead of warm fronts. Similar ascent is seen around tropical cyclones outside of the eyewall, and in comma-head precipitation patterns around mid-latitude cyclones. A wide variety of weather can be found along an occluded front, with thunderstorms possible, but usually their passage is associated with a drying of the air mass. Occluded fronts usually form around mature low-pressure areas. Precipitation may occur on celestial bodies other than Earth. When it gets cold, Mars has precipitation that most likely takes the form of ice needles, rather than rain or snow.

Convection

Convective rain, or showery precipitation, occurs from convective clouds, e.g., cumulonimbus or cumulus congestus. It falls as showers with rapidly changing intensity. Convective precipitation falls over a certain area for a relatively short time, as convective clouds have limited horizontal extent. Most precipitation in the tropics appears to be convective; however, it has been suggested that stratiform precipitation also occurs. Graupel and hail indicate convection. In mid-latitudes, convective precipitation is intermittent and often associated with baroclinic boundaries such as cold fronts, squall lines, and warm fronts.

Convective precipitation

Orographic Effects

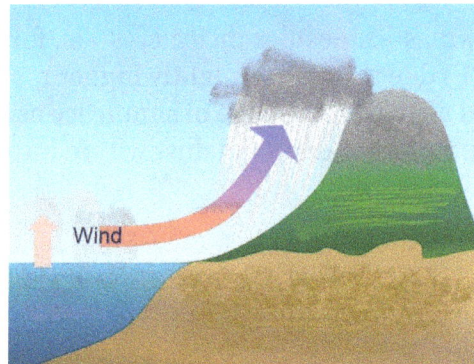

Orographic precipitation

Orographic precipitation occurs on the windward side of mountains and is caused by the rising air motion of a large-scale flow of moist air across the mountain ridge, resulting in adiabatic cooling and condensation. In mountainous parts of the world subjected to relatively consistent winds (for example, the trade winds), a more moist climate usually prevails on the windward side of a mountain than on the leeward or downwind side. Moisture is removed by orographic lift, leaving drier air on the descending and generally warming, leeward side where a rain shadow is observed.

In Hawaii, Mount Wai'ale'ale, on the island of Kauai, is notable for its extreme rainfall, as it has the second highest average annual rainfall on Earth, with 12,000 millimetres (460 in). Storm systems affect the state with heavy rains between October and March. Local climates vary considerably on each island due to their topography, divisible into windward and leeward regions based upon location relative to the higher mountains. Windward sides face the east to northeast trade winds and receive much more rainfall; leeward sides are drier and sunnier, with less rain and less cloud cover.

In South America, the Andes mountain range blocks Pacific moisture that arrives in that continent, resulting in a desertlike climate just downwind across western Argentina.

The Sierra Nevada range creates the same effect in North America forming the Great Basin and Mojave Deserts. Similarly, in Asia, the Himalaya mountains create an obstacle to monsoons which leads to extremely high precipitation on the southern side and lower preciptation levels on the northern side.

Snow

Lake-effect snow bands near the Korean Peninsula

Extratropical cyclones can bring cold and dangerous conditions with heavy rain and snow with winds exceeding 119 km/h (74 mph), (sometimes referred to as windstorms in Europe). The band of precipitation that is associated with their warm front is often extensive, forced by weak upward vertical motion of air over the frontal boundary which condenses as it cools and produces precipitation within an elongated band, which is wide and stratiform, meaning falling out of nimbostratus clouds. When moist air tries to dislodge an arctic air mass, overrunning snow can result within the poleward side of the elongated precipitation band. In the Northern Hemisphere, poleward is towards the North Pole, or north. Within the Southern Hemisphere, poleward is towards the South Pole, or south.

Southwest of extratropical cyclones, curved cyclonic flow bringing cold air across the relatively warm water bodies can lead to narrow lake-effect snow bands. Those bands bring strong localized snowfall which can be understood as follows: Large water bodies such as lakes efficiently store heat that results in significant temperature differences (larger than 13 °C or 23 °F) between the water surface and the air above. Because of this temperature difference, warmth and moisture are transported upward, condensing into vertically oriented clouds which produce snow showers. The temperature decrease with height and cloud depth are directly affected by both the water temperature and the large-scale environment. The stronger the temperature decrease with height, the deeper the clouds get, and the greater the precipitation rate becomes.

In mountainous areas, heavy snowfall accumulates when air is forced to ascend the mountains and squeeze out precipitation along their windward slopes, which in cold conditions, falls in the form of snow. Because of the ruggedness of terrain, forecasting the location of heavy snowfall remains a significant challenge.

Within the Tropics

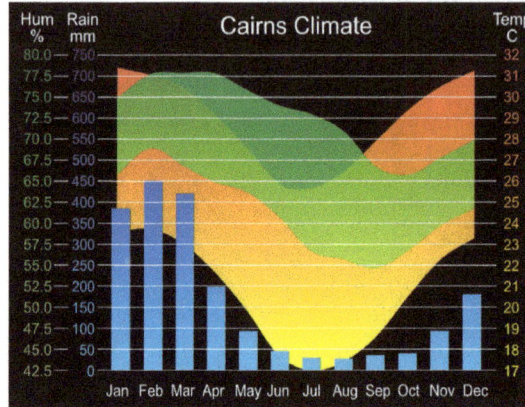

Rainfall distribution by month in Cairns showing the extent of the wet season at that location

The wet, or rainy, season is the time of year, covering one or more months, when most of the average annual rainfall in a region falls. The term *green season* is also sometimes used as a euphemism by tourist authorities. Areas with wet seasons are dispersed across portions of the tropics and subtropics. Savanna climates and areas with monsoon regimes have wet summers and dry winters. Tropical rainforests technically do not have dry or wet seasons, since their rainfall is equally distributed through the year. Some areas with pronounced rainy seasons will see a break in rainfall mid-season when the intertropical convergence zone or monsoon trough move poleward of their location during the middle of the warm season. When the wet season occurs during the warm season, or summer, rain falls mainly during the late afternoon and early evening hours. The wet season is a time when air quality improves, freshwater quality improves, and vegetation grows significantly. Soil nutrients diminish and erosion increases. Animals have adaptation and survival strategies for the wetter regime. The previous dry season leads to food shortages into the wet season, as the crops have yet to mature. Developing countries have noted that their populations show seasonal weight fluctuations due to food shortages seen before the first harvest, which occurs late in the wet season.

Tropical cyclones, a source of very heavy rainfall, consist of large air masses several hundred miles across with low pressure at the centre and with winds blowing inward towards the centre in either a clockwise direction (southern hemisphere) or counterclockwise (northern hemisphere). Although cyclones can take an enormous toll in lives and personal property, they may be important factors in the precipitation regimes of places they impact, as they may bring much-needed precipitation to otherwise dry regions. Areas in their path can receive a year's worth of rainfall from a tropical cyclone passage.

Large-scale Geographical Distribution

On the large scale, the highest precipitation amounts outside topography fall in the tropics, closely tied to the Intertropical Convergence Zone, itself the ascending branch of the Hadley cell. Mountainous locales near the equator in Colombia are amongst the wettest places on Earth. North and south of this are regions of descending air that form subtropical ridges where precipitation is low; the land surface underneath is usually arid, which forms most of the Earth's deserts. An exception to this rule is in Hawaii, where upslope flow due to the trade winds lead to one of the wettest locations on Earth. Otherwise, the flow of the Westerlies into the Rocky Mountains lead to the wettest, and at elevation snowiest, locations within North America. In Asia during the wet season, the flow of moist air into the Himalayas leads to some of the greatest rainfall amounts measured on Earth in northeast India.

Measurement

Standard rain gauge

The standard way of measuring rainfall or snowfall is the standard rain gauge, which can be found in 100 mm (4 in) plastic and 200 mm (8 in) metal varieties. The inner cylinder is filled by 25 mm (1 in) of rain, with overflow flowing into the outer cylinder. Plastic gauges have markings on the inner cylinder down to 0.25 mm (0.01 in) resolution, while metal gauges require use of a stick designed with the appropriate 0.25 mm (0.01 in) markings. After the inner cylinder is filled, the amount inside it is discarded, then filled with the remaining rainfall in the outer cylinder until all the fluid in the outer cylinder is gone, adding to the overall total until the outer cylinder is empty. These gauges are used in the winter by removing the funnel and inner cylinder and allowing snow and freezing rain to collect inside the outer cylinder. Some add anti-freeze to their gauge so they do not have to melt the snow or ice that falls into the gauge. Once the snowfall/ice is finished accumulating, or as 300 mm (12 in) is approached, one can either bring it inside to melt, or use lukewarm water to fill the inner cylinder with in order to melt the frozen precipitation in the outer cylinder, keeping track of the warm fluid added, which is subsequently subtracted from the overall total once all the ice/snow is melted.

Other types of gauges include the popular wedge gauge (the cheapest rain gauge and most fragile), the tipping bucket rain gauge, and the weighing rain gauge. The wedge and tipping bucket gauges will have problems with snow. Attempts to compensate for snow/ice by warming the tipping bucket meet with limited success, since snow may sublimate if the gauge is kept much above freezing. Weighing gauges with antifreeze should do fine with snow, but again, the funnel needs to be removed before the event begins. For those looking to measure rainfall the most inexpensively, a can that is cylindrical with straight sides will act as a rain gauge if left out in the open, but its accuracy will depend on what ruler is used to measure the rain with. Any of the above rain gauges can be made at home, with enough know-how.

When a precipitation measurement is made, various networks exist across the United States and elsewhere where rainfall measurements can be submitted through the Internet, such as CoCoRAHS or GLOBE. If a network is not available in the area where one lives, the nearest local weather office will likely be interested in the measurement.

Hydrometeor Definition

A concept used in precipitation measurement is the hydrometeor. Bits of liquid or solid water in the atmosphere are known as hydrometeors. Formations due to condensation, such as clouds, haze, fog, and mist, are composed of hydrometeors. All precipitation types are made up of hydrometeors by definition, including virga, which is precipitation which evaporates before reaching the ground. Particles blown from the Earth's surface by wind, such as blowing snow and blowing sea spray, are also hydrometeors.

Satellite Estimates

Although surface precipitation gauges are considered the standard for measuring precipitation, there are many areas in which their use is not feasible. This includes the vast expanses of ocean and remote land areas. In other cases, social, technical or administrative issues prevent the dissemination of gauge observations. As a result, the modern global record of precipitation largely depends on satellite observations.

Satellite sensors work by remotely sensing precipitation—recording various parts of the electromagnetic spectrum that theory and practice show are related to the occurrence and intensity of precipitation. The sensors are almost exclusively passive, recording what they see, similar to a camera, in contrast to active sensors (radar, lidar) that send out a signal and detect its impact on the area being observed.

Satellite sensors now in practical use for precipitation fall into two categories. Thermal infrared (IR) sensors record a channel around 11 micron wavelength and primarily give information about cloud tops. Due to the typical structure of the atmosphere, cloud-top temperatures are approximately inversely related to cloud-top heights, meaning colder clouds almost always occur at higher altitudes. Further, cloud tops with a lot of small-scale variation are likely to be more vigorous than smooth-topped clouds. Various mathematical schemes, or algorithms, use these and other properties to estimate precipitation from the IR data.

The second category of sensor channels is in the microwave part of the electromagnetic spectrum. The frequencies in use range from about 10 gigahertz to a few hundred GHz. Channels up to about 37 GHz primarily provide information on the liquid hydrometeors (rain and drizzle) in the lower parts of clouds, with larger amounts of liquid emitting higher amounts of microwave radiant energy. Channels above 37 GHz display emission signals, but are dominated by the action of solid hydrometeors (snow, graupel, etc.) to scatter microwave radiant energy. Satellites such as the Tropical Rainfall Measuring Mission (TRMM) and the Global Precipitation Measurement (GPM) mission employ microwave sensors to form precipitation estimates.

Additional sensor channels and products have been demonstrated to provide additional useful information including visible channels, additional IR channels, water vapor channels and atmospheric sounding retrievals. However, most precipitation data sets in current use do not employ these data sources.

Satellite Data Sets

The IR estimates have rather low skill at short time and space scales, but are available very frequently (15 minutes or more often) from satellites in geosynchronous Earth orbit. IR works best in cases of deep, vigorous convection—such as the tropics—and becomes progressively less useful in areas where stratiform (layered) precipitation dominates, especially in mid- and high-latitude regions. The more-direct physical connection between hydrometeors and microwave channels gives the microwave estimates greater skill on short time and space scales than is true for IR. However, microwave sensors fly only on low Earth orbit satellites, and there are few enough of them that the average time between observations exceeds three hours. This several-hour interval is insufficient to adequately document precipitation because of the transient nature of most precipitation systems as well as the inability of a single satellite to appropriately capture the typical daily cycle of precipitation at a given location.

Since the late 1990s, several algorithms have been developed to combine precipitation data from multiple satellites' sensors, seeking to emphasize the strengths and minimize the weaknesses of the individual input data sets. The goal is to provide "best" estimates of precipitation on a uniform time/space grid, usually for as much of the globe as possible. In some cases the long-term homogeneity of the dataset is emphasized, which is the Climate Data Record standard.

In other cases, the goal is producing the best instantaneous satellite estimate, which is the High Resolution Precipitation Product approach. In either case, of course, the less-emphasized goal is also considered desirable. One key result of the multi-satellite studies is that including even a small amount of surface gauge data is very useful for controlling the biases that are endemic to satellite estimates. The difficulties in using gauge data are that 1) their availability is limited, as noted above, and 2) the best analyses of gauge data take two months or more after the observation time to undergo the necessary transmission, assembly, processing and quality control. Thus, precipitation estimates that include gauge data tend to be produced further after the observation time than the no-gauge estimates. As a result, while estimates that include gauge data may provide a more accurate depiction of the "true" precipitation, they are generally not suited for real- or near-real-time applications.

The work described has resulted in a variety of datasets possessing different formats, time/space grids, periods of record and regions of coverage, input datasets, and analysis procedures, as well as many different forms of dataset version designators. In many cases, one of the modern multi-satellite data sets is the best choice for general use.

Return Period

The likelihood or probability of an event with a specified intensity and duration, is called the return period or frequency. The intensity of a storm can be predicted for any return period and storm duration, from charts based on historic data for the location. The term *1 in 10 year storm* describes a rainfall event which is rare and is only likely to occur once every 10 years, so it has a 10 percent likelihood any given year. The rainfall will be greater and the flooding will be worse than the worst storm expected in any single year. The term *1 in 100 year storm* describes a rainfall event which is extremely rare and which will occur with a likelihood of only once in a century, so has a 1 percent likelihood in any given year. The rainfall will be extreme and flooding to be worse than a 1 in 10 year event. As with all probability events, it is possible though unlikely to have two "1 in 100 Year Storms" in a single year.

Role in Köppen Climate Classification

The Köppen classification depends on average monthly values of temperature and precipitation. The most commonly used form of the Köppen classification has five primary types labeled A through E. Specifically, the primary types are A, tropical; B, dry; C, mild mid-latitude; D, cold mid-latitude; and E, polar. The five primary classifications can be further divided into secondary classifications such as rain forest, monsoon, tropical savanna, humid subtropical, humid continental, oceanic climate, Mediterranean climate, steppe, subarctic climate, tundra, polar ice cap, and desert.

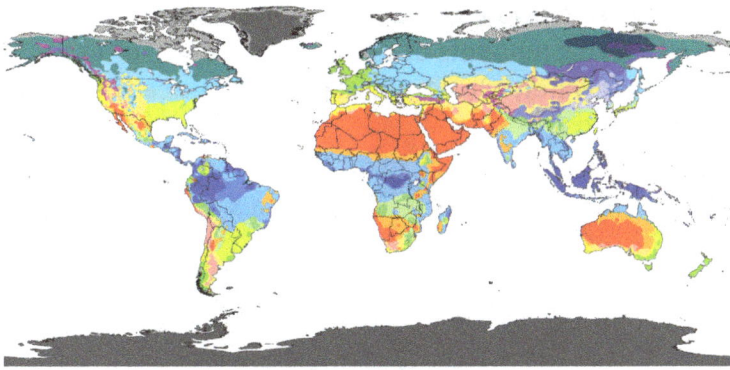

Updated Köppen-Geiger climate map

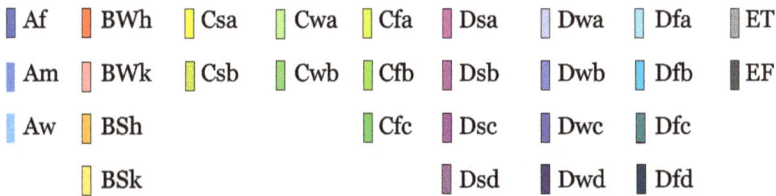

Af	BWh	Csa	Cwa	Cfa	Dsa	Dwa	Dfa	ET
Am	BWk	Csb	Cwb	Cfb	Dsb	Dwb	Dfb	EF
Aw	BSh		Cwc	Cfc	Dsc	Dwc	Dfc	
	BSk				Dsd	Dwd	Dfd	

Rain forests are characterized by high rainfall, with definitions setting minimum normal annual rainfall between 1,750 and 2,000 mm (69 and 79 in). A tropical savanna is a grassland biome located in semi-arid to semi-humid climate regions of subtropical and tropical latitudes, with rainfall between 750 and 1,270 mm (30 and 50 in) a year. They are widespread on Africa, and are also found in India, the northern parts of South America, Malaysia, and Australia. The humid subtropical climate zone is where winter rainfall (and sometimes snowfall) is associated with large storms that the westerlies steer from west to east. Most summer rainfall occurs during thunderstorms and from occasional tropical cyclones. Humid subtropical climates lie on the east side continents, roughly between latitudes 20° and 40° degrees away from the equator.

An oceanic (or maritime) climate is typically found along the west coasts at the middle latitudes of all the world's continents, bordering cool oceans, as well as southeastern Australia, and is accompanied by plentiful precipitation year-round. The Mediterranean climate regime resembles the climate of the lands in the Mediterranean Basin, parts of western North America, parts of Western and South Australia, in southwestern South Africa and in parts of central Chile. The climate is characterized by hot, dry summers and cool, wet winters. A steppe is a dry grassland. Subarctic climates are cold with continuous permafrost and little precipitation.

Effect on Agriculture

Precipitation, especially rain, has a dramatic effect on agriculture. All plants need at least some water to survive, therefore rain (being the most effective means of watering) is important to agriculture. While a regular rain pattern is usually vital to healthy plants, too much or too little rainfall can be harmful, even devastating to crops. Drought can kill crops and increase erosion, while overly wet weather can cause harmful fungus growth. Plants need varying amounts of rainfall to survive. For example, certain cacti require small amounts of water, while tropical plants may need up to hundreds of inches of rain per year to survive.

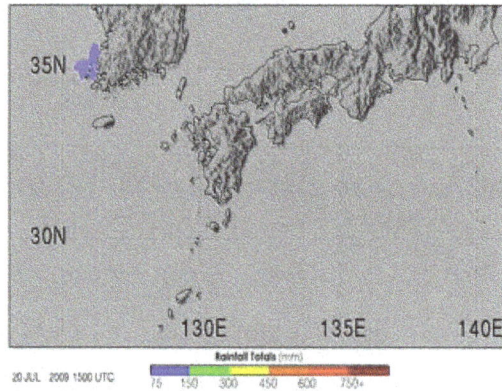

Rainfall estimates for southern Japan and the surrounding region from July 20 to 27, 2009.

In areas with wet and dry seasons, soil nutrients diminish and erosion increases during the wet season. Animals have adaptation and survival strategies for the wetter regime. The previous dry season leads to food shortages into the wet season, as the crops have yet to mature. Developing countries have noted that their populations show seasonal weight fluctuations due to food shortages seen before the first harvest, which occurs late in the wet season.

Changes due to Global Warming

Increasing temperatures tend to increase evaporation which leads to more precipitation. Precipitation has generally increased over land north of 30°N from 1900 to 2005 but has declined over the tropics since the 1970s. Globally there has been no statistically significant overall trend in precipitation over the past century, although trends have varied widely by region and over time. Eastern portions of North and South America, northern Europe, and northern and central Asia have become wetter. The Sahel, the Mediterranean, southern Africa and parts of southern Asia have become drier. There has been an increase in the number of heavy precipitation events over many areas during the past century, as well as an increase since the 1970s in the prevalence of droughts—especially in the tropics and subtropics. Changes in precipitation and evaporation over the oceans are suggested by the decreased salinity of mid- and high-latitude waters (implying more precipitation), along with increased salinity in lower latitudes (implying less precipitation, more evaporation, or both). Over the contiguous United States, total annual precipitation increased at an average rate of 6.1% per century since 1900, with the greatest increases within the East North Central climate region (11.6% per century) and the South (11.1%). Hawaii was the only region to show a decrease (-9.25%).

Changes due to Urban Heat Island

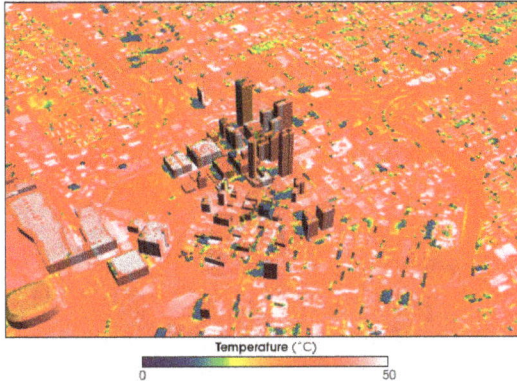

Temperature (°C)
0   50

Image of Atlanta, Georgia, showing temperature distribution, with hot areas appearing white

The urban heat island warms cities 0.6 to 5.6 °C (1.1 to 10.1 °F) above surrounding suburbs and rural areas. This extra heat leads to greater upward motion, which can induce additional shower and thunderstorm activity. Rainfall rates downwind of cities are increased between 48% and 116%. Partly as a result of this warming, monthly rainfall is about 28% greater between 32 to 64 kilometres (20 to 40 mi) downwind of cities, compared with upwind. Some cities induce a total precipitation increase of 51%.

Forecasting

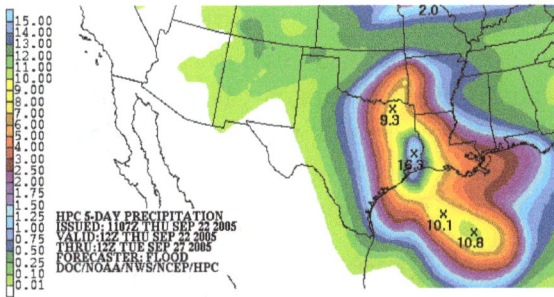

Example of a five-day rainfall forecast from the Hydrometeorological Prediction Center

The Quantitative Precipitation Forecast (QPF) is the expected amount of liquid precipitation accumulated over a specified time period over a specified area. A QPF will be specified when a measurable precipitation type reaching a minimum threshold is forecast for any hour during a QPF valid period. Precipitation forecasts tend to be bound by synoptic hours such as 0000, 0600, 1200 and 1800 GMT. Terrain is considered in QPFs by use of topography or based upon climatological precipitation patterns from observations with fine detail. Starting in the mid to late 1990s, QPFs were used within hydrologic forecast models to simulate impact to rivers throughout the United States. Forecast models show significant sensitivity to humidity levels within the planetary boundary layer, or in the lowest levels of the atmosphere, which decreases with height. QPF can be generated on a quantitative, forecasting amounts, or a qualitative, forecasting the probability of a specific amount, basis. Radar imagery forecasting techniques show higher skill than model forecasts within six to seven hours of the time of the radar image. The forecasts can be verified through use of rain gauge measurements, weather radar estimates, or a combination of both. Various skill scores can be determined to measure the value of the rainfall forecast.

Surface Runoff

Runoff flowing into a stormwater drain

Surface runoff (also known as overland flow) is the flow of water that occurs when excess stormwater, meltwater, or other sources flows over the Earth's surface. This might occur because soil is saturated to full capacity, because rain arrives more quickly than soil can absorb it, or because impervious areas (roofs and pavement) send their runoff to surrounding soil that cannot absorb all of it. Surface runoff is a major component of the water cycle. It is the primary agent in soil erosion by water.

Runoff that occurs on the ground surface before reaching a channel is also called a nonpoint source. If a nonpoint source contains man-made contaminants, or natural forms of pollution (such as rotting leaves) the runoff is called nonpoint source pollution. A land area which produces runoff that drains to a common point is called a drainage basin. When runoff flows along the ground, it can pick up soil contaminants including petroleum, pesticides, or fertilizers that become discharge or nonpoint source pollution.

In addition to causing water erosion and pollution, surface runoff in urban areas is a primary cause of urban flooding which can result in property damage, damp and mold in basements, and street flooding.

Generation

Surface runoff can be generated either by rainfall, snowfall or by the melting of snow, or glaciers.

Surface runoff from a hillside after soil is saturated

Snow and glacier melt occur only in areas cold enough for these to form permanently. Typically snowmelt will peak in the spring and glacier melt in the summer, leading to pronounced flow maxima in rivers affected by them. The determining factor of the rate of melting of snow or glaciers is both air temperature and the duration of sunlight. In high mountain regions, streams frequently rise on sunny days and fall on cloudy ones for this reason.

In areas where there is no snow, runoff will come from rainfall. However, not all rainfall will produce runoff because storage from soils can absorb light showers. On the extremely ancient soils of Australia and Southern Africa, proteoid roots with their extremely dense networks of root hairs can absorb so much rainwater as to prevent runoff even when substantial amounts of rain fall. In these regions, even on less infertile cracking clay soils, high amounts of rainfall and potential evaporation are needed to generate any surface runoff, leading to specialised adaptations to extremely variable (usually ephemeral) streams.

Infiltration Excess Overland Flow

This occurs when the rate of rainfall on a surface exceeds the rate at which water can infiltrate the ground, and any depression storage has already been filled. This is called flooding excess overland flow, Hortonian overland flow (after Robert E. Horton), or unsaturated overland flow. This more commonly occurs in arid and semi-arid regions, where rainfall intensities are high and the soil infiltration capacity is reduced because of surface sealing, or in paved areas. This occurs largely in city areas where pavements prevent water from flooding.

Saturation Excess Overland Flow

When the soil is saturated and the depression storage filled, and rain continues to fall, the rainfall will immediately produce surface runoff. The level of antecedent soil moisture is one factor affecting the time until soil becomes saturated. This runoff is called saturation excess overland flow or saturated overland flow.

Antecedent Soil Moisture

Soil retains a degree of moisture after a rainfall. This residual water moisture affects the soil's infiltration capacity. During the next rainfall event, the infiltration capacity will cause the soil to be saturated at a different rate. The higher the level of antecedent soil moisture, the more quickly the soil becomes saturated. Once the soil is saturated, runoff occurs.

Subsurface Return Flow

After water infiltrates the soil on an up-slope portion of a hill, the water may flow laterally through the soil, and exfiltrate (flow out of the soil) closer to a channel. This is called subsurface return flow or throughflow.

As it flows, the amount of runoff may be reduced in a number of possible ways: a small portion of it may evapotranspire; water may become temporarily stored in microtopographic depressions; and a portion of it may infiltrate as it flows overland. Any remaining surface water eventually flows into a receiving water body such as a river, lake, estuary or ocean.

Human Influence

Urban surface water runoff

Urbanization increases surface runoff by creating more impervious surfaces such as pavement and buildings that do not allow percolation of the water down through the soil to the aquifer. It is instead forced directly into streams or storm water runoff drains, where erosion and siltation can be major problems, even when flooding is not. Increased runoff reduces groundwater recharge, thus lowering the water table and making droughts worse, especially for farmers and others who depend on the water wells.

When anthropogenic contaminants are dissolved or suspended in runoff, the human impact is expanded to create water pollution. This pollutant load can reach various receiving waters such as streams, rivers, lakes, estuaries and oceans with resultant water chemistry changes to these water systems and their related ecosystems.

A 2008 report by the United States National Research Council identified urban storm-water as a leading source of water quality problems in the U.S.

As humans continue to alter the climate through the addition of greenhouse gases to the atmosphere, precipitation patterns are expected to change as the atmospheric capacity for water vapor increases. This will have direct consequences on runoff amounts.

Effects of Surface Runoff

Erosion and Deposition

Surface runoff can cause erosion of the Earth's surface; eroded material may be deposited a considerable distance away. There are four main types of soil erosion by water: splash erosion, sheet erosion, rill erosion and gully erosion. Splash erosion is the result of mechanical collision of raindrops with the soil surface: soil particles which are dislodged by the impact then move with the surface runoff. Sheet erosion is the overland transport of sediment by runoff without a well defined channel. Soil surface roughness causes may cause runoff to become concentrated into narrower flow paths: as these incise, the small but well-defined channels which are formed are known as rills. These channels can be as small as one centimeter wide or as large as several meters. If runoff continue to incise and enlarge rills, they may eventually grow to become gullies. Gully erosion can transport large amounts of eroded material in a small time period.

Soil erosion by water on intensively-tilled farmland.

Willow hedge strengthened with fascines for the limitation of runoff, north of France

Reduced crop productivity usually results from erosion, and these effects are studied in the field of soil conservation. The soil particles carried in runoff vary in size from about .001 millimeter to 1.0 millimeter in diameter. Larger particles settle over short transport distances, whereas small particles can be carried over long distances suspended in the water column. Erosion of silty soils that contain smaller particles generates turbidity and diminishes light transmission, which disrupts aquatic ecosystems.

Entire sections of countries have been rendered unproductive by erosion. On the high central plateau of Madagascar, approximately ten percent of that country's land area, virtually the entire landscape is devoid of vegetation, with erosive gully furrows typically in excess of 50 meters deep and one kilometer wide. Shifting cultivation is a farming system which sometimes incorporates the slash and burn method in some regions of the world. Erosion causes loss of the fertile top soil and reduces its fertility and quality of the agricultural produce.

Modern industrial farming is another major cause of erosion. In some areas in the American corn belt, more than 50 percent of the original topsoil has been carried away within the last 100 years.

Environmental Effects

The principal environmental issues associated with runoff are the impacts to surface water, groundwater and soil through transport of water pollutants to these systems. Ultimately these consequences translate into human health risk, ecosystem disturbance and aesthetic impact to water resources. Some of the contaminants that create the greatest impact to surface waters arising from runoff are petroleum substances, herbicides and fertilizers. Quantitative uptake by surface runoff of pesticides and other contaminants has been studied since the 1960s, and early on contact of pesticides with water was known to enhance phytotoxicity. In the case of surface waters, the impacts translate to water pollution, since the streams and rivers have received runoff carrying various chemicals or sediments. When surface waters are used as potable water supplies, they can be compromised regarding health risks and drinking water aesthetics (that is, odor, color and turbidity effects). Contaminated surface waters risk altering the metabolic processes of the aquatic species that they host; these alterations can lead to death, such as fish kills, or alter the balance of populations present. Other specific impacts are on animal mating, spawning, egg and larvae viability, juvenile survival and plant productivity. Some researches show surface runoff of pesticides, such as DDT, can alter the gender of fish species genetically, which transforms male into female fish.

Surface runoff occurring within forests can supply lakes with high loads of mineral nitrogen and phosphorus leading to eutrophication. Runoff waters within coniferous forests are also enriched with humic acids and can lead to humification of water bodies Additionally, high standing and young islands in the tropics and subtropics can undergo high soil erosion rates and also contribute large material fluxes to the coastal ocean. Such land derived runoff of sediment nutrients, carbon, and contaminants can have large impacts on global biogeochemical cycles and marine and coastal ecosystems.

In the case of groundwater, the main issue is contamination of drinking water, if the aquifer is abstracted for human use. Regarding soil contamination, runoff waters can have two important pathways of concern. Firstly, runoff water can extract soil contaminants and carry them in the form of water pollution to even more sensitive aquatic

habitats. Secondly, runoff can deposit contaminants on pristine soils, creating health or ecological consequences.

Agricultural Issues

The other context of agricultural issues involves the transport of agricultural chemicals (nitrates, phosphates, pesticides, herbicides etc.) via surface runoff. This result occurs when chemical use is excessive or poorly timed with respect to high precipitation. The resulting contaminated runoff represents not only a waste of agricultural chemicals, but also an environmental threat to downstream ecosystems.

Flooding

Flooding occurs when a watercourse is unable to convey the quantity of runoff flowing downstream. The frequency with which this occurs is described by a return period. Flooding is a natural process, which maintains ecosystem composition and processes, but it can also be altered by land use changes such as river engineering. Floods can be both beneficial to societies or cause damage. Agriculture along the Nile floodplain took advantage of the seasonal flooding that deposited nutrients beneficial for crops. However, as the number and susceptibility of settlements increase, flooding increasingly becomes a natural hazard. In urban areas, surface runoff is the primary cause of urban flooding, known for its repetitive and costly impact on communities. Adverse impacts span loss of life, property damage, contamination of water supplies, loss of crops, and social dislocation and temporary homelessness. Floods are among the most devastating of natural disasters.

Mitigation and Treatment

Runoff holding ponds (Uplands neighborhood of North Bend, Washington)

Mitigation of adverse impacts of runoff can take several forms:

- Land use development controls aimed at minimizing impervious surfaces in urban areas

- Erosion controls for farms and construction sites

- Flood control and retrofit programs, such as green infrastructure

- Chemical use and handling controls in agriculture, landscape maintenance, industrial use, etc.

Land use controls. Many world regulatory agencies have encouraged research on methods of minimizing total surface runoff by avoiding unnecessary hardscape. Many municipalities have produced guidelines and codes (zoning and related ordinances) for land developers that encourage minimum width sidewalks, use of pavers set in earth for driveways and walkways and other design techniques to allow maximum water infiltration in urban settings. An example land use control program can be seen in the city of Santa Monica, California.

Erosion controls have appeared since medieval times when farmers realized the importance of contour farming to protect soil resources. Beginning in the 1950s these agricultural methods became increasingly more sophisticated. In the 1960s some state and local governments began to focus their efforts on mitigation of construction runoff by requiring builders to implement erosion and sediment controls (ESCs). This included such techniques as: use of straw bales and barriers to slow runoff on slopes, installation of silt fences, programming construction for months that have less rainfall and minimizing extent and duration of exposed graded areas. Montgomery County, Maryland implemented the first local government sediment control program in 1965, and this was followed by a statewide program in Maryland in 1970.

Flood control programs as early as the first half of the twentieth century became quantitative in predicting peak flows of riverine systems. Progressively strategies have been developed to minimize peak flows and also to reduce channel velocities. Some of the techniques commonly applied are: provision of holding ponds (also called detention basins) to buffer riverine peak flows, use of energy dissipators in channels to reduce stream velocity and land use controls to minimize runoff.

Chemical use and handling. Following enactment of the U.S. Resource Conservation and Recovery Act (RCRA) in 1976, and later the Water Quality Act of 1987, states and cities have become more vigilant in controlling the containment and storage of toxic chemicals, thus preventing releases and leakage. Methods commonly applied are: requirements for double containment of underground storage tanks, registration of hazardous materials usage, reduction in numbers of allowed pesticides and more stringent regulation of fertilizers and herbicides in landscape maintenance. In many industrial cases, pretreatment of wastes is required, to minimize escape of pollutants into sanitary or stormwater sewers.

The U.S. Clean Water Act (CWA) requires that local governments in urbanized areas (as defined by the Census Bureau) obtain stormwater discharge permits for their drainage systems. Essentially this means that the locality must operate a stormwater management program for all surface runoff that enters the municipal separate storm sewer system ("MS4"). EPA and state regulations and related publications outline six basic components that each local program must contain:

- Public education (informing individuals, households, businesses about ways to avoid stormwater pollution)

- Public involvement (support public participation in implementation of local programs)

- Illicit discharge detection & elimination (removing sanitary sewer or other non-stormwater connections to the MS4)

- Construction site runoff controls (i.e. erosion & sediment controls)

- Post-construction (i.e. permanent) stormwater management controls

- Pollution prevention and "good housekeeping" measures (e.g. system maintenance).

Other property owners which operate storm drain systems similar to municipalities, such as state highway systems, universities, military bases and prisons, are also subject to the MS4 permit requirements.

Measurement and Mathematical Modeling

Runoff is analyzed by using mathematical models in combination with various water quality sampling methods. Measurements can be made using continuous automated water quality analysis instruments targeted on pollutants such as specific organic or inorganic chemicals, pH, turbidity etc. or targeted on secondary indicators such as dissolved oxygen. Measurements can also be made in batch form by extracting a single water sample and conducting any number of chemical or physical tests on that sample.

In the 1950s or earlier hydrology transport models appeared to calculate quantities of runoff, primarily for flood forecasting. Beginning in the early 1970s computer models were developed to analyze the transport of runoff carrying water pollutants, which considered dissolution rates of various chemicals, infiltration into soils and ultimate pollutant load delivered to receiving waters. One of the earliest models addressing chemical dissolution in runoff and resulting transport was developed in the early 1970s under contract to the United States Environmental Protection Agency (EPA). This computer model formed the basis of much of the mitigation study that led to strategies for land use and chemical handling controls.

Other computer models have been developed (such as the DSSAM Model) that allow surface runoff to be tracked through a river course as reactive water pollutants. In this case the surface runoff may be considered to be a line source of water pollution to the receiving waters.

Storage

Portion of the precipitation falling on land surface which does not flow out as runoff gets stored as either as surface water bodies like Lakes, Reservoirs and Wetlands or as sub-surface water body, usually called Ground water.

Ground water storage is the water infiltrating through the soil cover of a land surface and traveling further to reach the huge body of water underground. As mentioned earlier, the amount of ground water storage is much greater than that of lakes and rivers. However, it is not possible to extract the entire groundwater by practicable means. It is interesting to note that the groundwater also is in a state of continuous movement – flowing from regions of higher potential to lower. The rate of movement, however, is exceptionally small compared to the surface water movement.

The following definitions may be useful:

Lakes: Large, naturally occurring inland body of water

Reservoirs: Artificial or natural inland body of water used to store water to meet various demands.

Wet Lands: Natural or artificial areas of shallow water or saturated soils that contain or could support water–loving plants.

Evapotranspiration

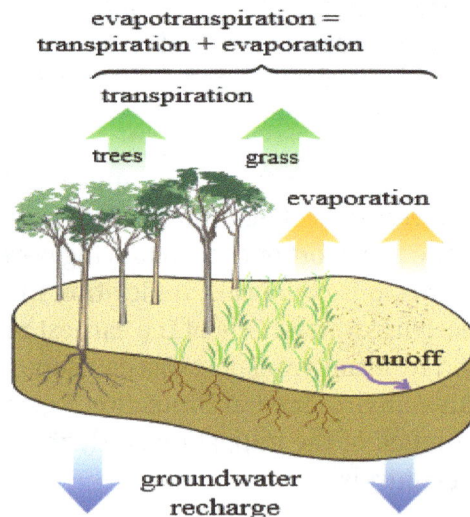

Water cycle of the Earth's surface, showing the individual components of transpiration and evaporation that make up evapotranspiration. Other closely related processes shown are runoff and groundwater recharge.

Evapotranspiration (ET) is the sum of evaporation and plant transpiration from the Earth's land and ocean surface to the atmosphere. Evaporation accounts for the movement of water to the air from sources such as the soil, canopy interception, and waterbodies. Transpiration accounts for the movement of water within a plant and the subsequent loss of water as vapor through stomata in its leaves. Evapotranspiration is an important part of the water cycle. An element (such as a tree) that contributes to evapotranspiration can be called an evapotranspirator.

Reference evapotranspiration (ET_o), sometimes incorrectly referred to as potential ET, is a representation of the environmental demand for evapotranspiration and represents the evapotranspiration rate of a short green crop (grass), completely shading the ground, of uniform height and with adequate water status in the soil profile. It is a reflection of the energy available to evaporate water, and of the wind available to transport the water vapour from the ground up into the lower atmosphere. Actual evapotranspiration is said to equal reference evapotranspiration when there is ample water. Some US states utilize a full cover alfalfa reference crop that is 0.5 m in height, rather than the short green grass reference, due to the higher value of ET from the alfalfa reference.

Water Cycle

Evapotranspiration is a significant water loss from drainage basins. Types of vegetation and land use significantly affect evapotranspiration, and therefore the amount of water leaving a drainage basin. Because water transpired through leaves comes from the roots, plants with deep reaching roots can more constantly transpire water. Herbaceous plants generally transpire less than woody plants because they usually have less extensive foliage. Conifer forests tend to have higher rates of evapotranspiration than deciduous forests, particularly in the dormant and early spring seasons. This is primarily due to the enhanced amount of precipitation intercepted and evaporated by conifer foliage during these periods. Factors that affect evapotranspiration include the plant's growth stage or level of maturity, percentage of soil cover, solar radiation, humidity, temperature, and wind. Isotope measurements indicate transpiration is the larger component of evapotranspiration.

Through evapotranspiration, forests reduce water yield, except in unique ecosystems called cloud forests. Trees in cloud forests collect the liquid water in fog or low clouds onto their surface, which drips down to the ground. These trees still contribute to evapotranspiration, but often collect more water than they evaporate or transpire.

In areas that are not irrigated, actual evapotranspiration is usually no greater than precipitation, with some buffer in time depending on the soil's ability to hold water. It will usually be less because some water will be lost due to percolation or surface runoff. An exception is areas with high water tables, where capillary action can cause water from the groundwater to rise through the soil matrix to the surface. If potential evapotranspiration is greater than actual precipitation, then soil will dry out, unless irrigation is used.

Evapotranspiration can never be greater than PET, but can be lower if there is not enough water to be evaporated or plants are unable to transpire readily.

Estimating Evapotranspiration

Evapotranspiration can be measured or estimated using several methods.

Indirect Methods

Pan evaporation data can be used to estimate lake evaporation, but transpiration and evaporation of intercepted rain on vegetation are unknown. There are three general approaches to estimate evapotranspiration indirectly.

Catchment Water Balance

Evapotranspiration may be estimated by creating an equation of the water balance of a drainage basin. The equation balances the change in water stored within the basin (S) with inputs and outgoes:

$$\Delta S = P - ET - Q - D$$

The input is precipitation (P) and the outgoes are evapotranspiration (which is to be estimated), streamflow (Q), and groundwater recharge (D). If the change in storage, precipitation, streamflow, and groundwater recharge are all estimated, the missing flux, ET, can be estimated by rearranging the above equation as follows:

$$ET = P - \Delta S - Q - D$$

Evapotranspiration of Urban Vegetation

The water requirement of urban landscapes, particularly urban parklands, is of growing concern. The estimation of evapotranspiration (ET) and subsequently plant water requirements in urban vegetation needs to consider the heterogeneity of plants, soils, water, and climate characteristics. In a research in South Australia, two practical ET estimation approaches are compared to a detailed Soil Water Balance (SWB) analysis over a one-year period. One approach is the Water Use Classification of Landscape Plants (WUCOLS) method, which is based on expert opinion on the water needs of different classes of landscape plants. The other is a remote sensing approach based on the Enhanced Vegetation Index (EVI) from Moderate Resolution Imaging Spectroradiometer (MODIS) sensors on the Terra satellite. Both methods require knowledge of reference ET calculated from meteorological data.

Hydrometeorological Equations

The most general and widely used equation for calculating reference ET is the Penman equation. The Penman-Monteith variation is recommended by the Food and Agriculture Organization and the American Society of Civil Engineers. The simpler Blaney-Criddle equation was popular in the Western United States for many years but it is not as accurate in regions with higher humidities. Other solutions used includes Makkink, which is simple but must be calibrated to a specific location, and Hargreaves. To convert the reference evapotranspiration to actual crop evapotranspiration, a crop coefficient and a stress coefficient must be used. Crop coefficients referred to in many hydrological models are themselves during periods for which the model is used. This is because crops are seasonal, perennial plants mature over multiple seasons, and stress responses can significantly depend upon many aspects of plant condition.

Energy Balance

A third methodology to estimate the actual evapotranspiration is the use of the energy balance.

$$\lambda E = R_n - G - H$$

where λE is the energy needed to change the phase of water from liquid to gas, R_n is the net radiation, G is the soil heat flux and H is the sensible heat flux. Using instruments like a scintillometer, soil heat flux plates or radiation meters, the components of the energy balance can be calculated and the energy available for actual evapotranspiration can be solved.

The SEBAL and METRIC algorithms solve the energy balance at the earth's surface using satellite imagery. This allows for both actual and potential evapotranspiration to be calculated on a pixel-by-pixel basis. Evapotranspiration is a key indicator for water management and irrigation performance. SEBAL and METRIC can map these key indicators in time and space, for days, weeks or years.

Experimental methods for Measuring Evapotranspiration

One method for measuring evapotranspiration is with a weighing lysimeter. The weight of a soil column is measured continuously and the change in storage of water in the soil is modeled by the change in weight. The change in weight is converted to units of length using the surface area of the weighing lysimeter and the unit weight of water. evapotranspiration is computed as the change in weight plus rainfall minus percolation.

Remote Sensing

In recent decades, estimating evapotranspiration has been improved by advances in remote sensing, particularly in agricultural studies. However, quantifying evapotranspiration from mixed vegetation environs, particularly urban parklands, is still challenging because of the heterogeneity of plant species, canopy covers and microclimates and because the methodology is costly. Different remote sensing-based approaches for estimating evapotranspiration have various advantages and disadvantages.

Eddy Covariance

The most direct method of measuring evapotranspiration is with the eddy covariance technique in which fast fluctuations of vertical wind speed are correlated with fast fluctuations in atmospheric water vapor density. This directly estimates the transfer of water vapor (evapotranspiration) from the land (or canopy) surface to the atmosphere.

Urban Landscape Plants

Methods for measuring evapotranspiration can be adapted to an urban setting to estimate the water requirements of urban landscape vegetation.

Potential Evapotranspiration

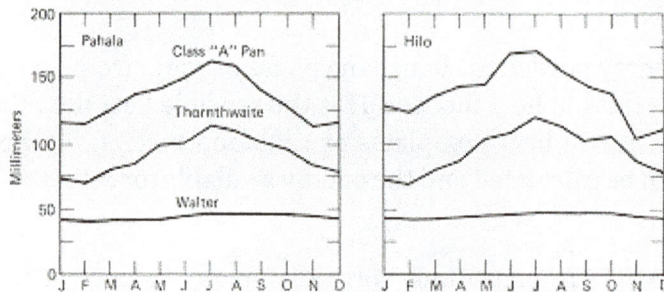

Monthly estimated potential evapotranspiration and measured pan evaporation for two locations in Hawaii, Hilo and Pahala.

Potential evapotranspiration (PET) is the amount of water that would be evaporated and transpired if there were sufficient water available. This demand incorporates the energy available for evaporation and the ability of the lower atmosphere to transport evaporated moisture away from the land surface. Potential evapotranspiration is higher in the summer, on less cloudy days, and closer to the equator, because of the higher levels of solar radiation that provides the energy for evaporation. Potential evapotranspiration is also higher on windy days because the evaporated moisture can be quickly moved from the ground or plant surface, allowing more evaporation to fill its place.

Potential evapotranspiration is expressed in terms of a depth of water, and can be graphed during the year.

Potential evapotranspiration is usually measured indirectly, from other climatic factors, but also depends on the surface type, such as free water (for lakes and oceans), the soil type for bare soil, and the vegetation. Often a value for the potential evapotranspiration is calculated at a nearby climate station on a reference surface, conventionally short grass. This value is called the reference evapotranspiration, and can be converted to a potential evapotranspiration by multiplying with a surface coefficient. In agriculture, this is called a crop coefficient. The difference between potential evapotranspiration and precipitation is used in irrigation scheduling.

Average annual potential evapotranspiration is often compared to average annual precipitation, P. The ratio of the two, P/PET, is the aridity index.

Water Resources

Where is Earth's Water?

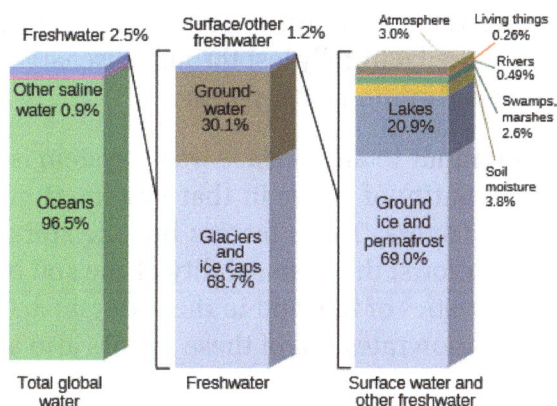

A graphical distribution of the locations of water on Earth. Only 3% of the Earth's water is fresh water. Most of it is in icecaps and glaciers (69%) and groundwater (30%), while all lakes, rivers and swamps combined only account for a small fraction (0.3%) of the Earth's total freshwater reserves.

Water resources are sources of water that are potentially useful. Uses of water include agricultural, industrial, household, recreational and environmental activities. The majority of human uses require fresh water.

97% of the water on the Earth is salt water and only three percent is fresh water; slightly over two thirds of this is frozen in glaciers and polar ice caps. The remaining unfrozen freshwater is found mainly as groundwater, with only a small fraction present above ground or in the air.

Fresh water is a renewable resource, yet the world's supply of groundwater is steadily decreasing, with depletion occurring most prominently in Asia, South America and North America, although it is still unclear how much natural renewal balances this usage, and whether ecosystems are threatened. The framework for allocating water resources to water users (where such a framework exists) is known as water rights.

Sources of Fresh Water

Surface Water

Lake Chungará and Parinacota volcano in northern Chile

Surface water is water in a river, lake or fresh water wetland. Surface water is naturally replenished by precipitation and naturally lost through discharge to the oceans, evaporation, evapotranspiration and groundwater recharge.

Although the only natural input to any surface water system is precipitation within its watershed, the total quantity of water in that system at any given time is also dependent on many other factors. These factors include storage capacity in lakes, wetlands and artificial reservoirs, the permeability of the soil beneath these storage bodies, the runoff characteristics of the land in the watershed, the timing of the precipitation and local evaporation rates. All of these factors also affect the proportions of water loss.

Human activities can have a large and sometimes devastating impact on these factors. Humans often increase storage capacity by constructing reservoirs and decrease it by draining wetlands. Humans often increase runoff quantities and velocities by paving areas and channelizing the stream flow.

The total quantity of water available at any given time is an important consideration. Some human water users have an intermittent need for water. For example, many farms require large quantities of water in the spring, and no water at all in the winter. To supply such a farm with water, a surface water system may require a large storage capacity to collect water throughout the year and release it in a short period of time. Other users have a continuous need for water, such as a power plant that requires water for cooling. To supply such a power plant with water, a surface water system only needs enough storage capacity to fill in when average stream flow is below the power plant's need.

Nevertheless, over the long term the average rate of precipitation within a watershed is the upper bound for average consumption of natural surface water from that watershed.

Natural surface water can be augmented by importing surface water from another watershed through a canal or pipeline. It can also be artificially augmented from any of the other sources listed here, however in practice the quantities are negligible. Humans can also cause surface water to be "lost" (i.e. become unusable) through pollution.

Brazil is the country estimated to have the largest supply of fresh water in the world, followed by Russia and Canada.

Panorama of a natural wetland (Sinclair Wetlands, New Zealand)

Under River Flow

Throughout the course of a river, the total volume of water transported downstream will often be a combination of the visible free water flow together with a substantial contribution flowing through rocks and sediments that underlie the river and its floodplain called the hyporheic zone. For many rivers in large valleys, this unseen component of flow may greatly exceed the visible flow. The hyporheic zone often forms a dynamic interface between surface water and groundwater from aquifers, exchanging flow between rivers and aquifers that may be fully charged or depleted. This is especially significant in karst areas where pot-holes and underground rivers are common.

Groundwater

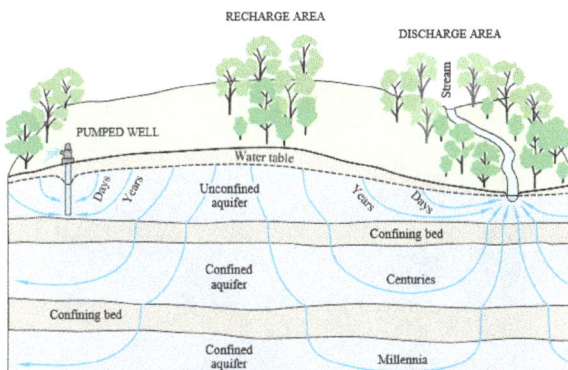

Relative groundwater travel times in the subsurface

Groundwater is fresh water located in the subsurface pore space of soil and rocks. It is also water that is flowing within aquifers below the water table. Sometimes it is useful to make a distinction between groundwater that is closely associated with surface water and deep groundwater in an aquifer (sometimes called "fossil water").

A shipot is a common water source in Central Ukrainian villages

Groundwater can be thought of in the same terms as surface water: inputs, outputs and storage. The critical difference is that due to its slow rate of turnover, groundwater storage is generally much larger (in volume) compared to inputs than it is for surface water. This difference makes it easy for humans to use groundwater unsustainably for a long time without severe consequences. Nevertheless, over the long term the average rate of seepage above a groundwater source is the upper bound for average consumption of water from that source.

The natural input to groundwater is seepage from surface water. The natural outputs from groundwater are springs and seepage to the oceans.

If the surface water source is also subject to substantial evaporation, a groundwater source may become saline. This situation can occur naturally under endorheic bodies of water, or artificially under irrigated farmland. In coastal areas, human use of a groundwater source may cause the direction of seepage to ocean to reverse which can also cause soil salinization. Humans can also cause groundwater to be "lost" (i.e. become unusable) through pollution. Humans can increase the input to a groundwater source by building reservoirs or detention ponds.

Frozen Water

Iceberg near Newfoundland

Several schemes have been proposed to make use of icebergs as a water source, however to date this has only been done for research purposes. Glacier runoff is considered to be surface water.

The Himalayas, which are often called "The Roof of the World", contain some of the most extensive and rough high altitude areas on Earth as well as the greatest area of glaciers and permafrost outside of the poles. Ten of Asia's largest rivers flow from there, and more than a billion people's livelihoods depend on them. To complicate matters, temperatures there are rising more rapidly than the global average. In Nepal, the temperature has risen by 0.6 degrees Celsius over the last decade, whereas globally, the Earth has warmed approximately 0.7 degrees Celsius over the last hundred years.

Desalination

Desalination is an artificial process by which saline water (generally sea water) is converted to fresh water. The most common desalination processes are distillation and reverse osmosis. Desalination is currently expensive compared to most alternative sources of water, and only a very small fraction of total human use is satisfied by desalination. It is only economically practical for high-valued uses (such as household and industrial uses) in arid areas. The most extensive use is in the Persian Gulf.

Water Uses

Agricultural

It is estimated that 70% of worldwide water is used for irrigation, with 15-35% of irrigation withdrawals being unsustainable. It takes around 2,000 - 3,000 litres of water to produce enough food to satisfy one person's daily dietary need. This is a considerable amount, when compared to that required for drinking, which is between two and five litres. To produce food for the now over 7 billion people who inhabit the planet today requires the water that would fill a canal ten metres deep, 100 metres wide and 2100 kilometres long.

Increasing Water Scarcity

Around fifty years ago, the common perception was that water was an infinite resource. At that time, there were fewer than half the current number of people on the planet. People were not as wealthy as today, consumed fewer calories and ate less meat, so less water was needed to produce their food. They required a third of the volume of water we presently take from rivers. Today, the competition for water resources is much more intense. This is because there are now seven billion people on the planet, their consumption of water-thirsty meat and vegetables is rising, and there is increasing competition for water from industry, urbanisation biofuel crops, and water reliant food items. In the future, even more water will be needed to produce food because the Earth's population is forecast to rise to 9 billion by 2050. An additional 2.5 or 3 billion people, choosing to eat fewer cereals and more meat and vegetables could add an additional five million kilometres to the virtual canal mentioned above.

An assessment of water management in agriculture sector was conducted in 2007 by the International Water Management Institute in Sri Lanka to see if the world had sufficient water to provide food for its growing population. It assessed the current availability of water for agriculture on a global scale and mapped out locations suffering from water scarcity. It found that a fifth of the world's people, more than 1.2 billion, live in areas of physical water scarcity, where there is not enough water to meet all demands. One third of the world's population does not have access to clean drinking water, which is more than 2.3 billion people. A further 1.6 billion people live in areas experiencing economic water scarcity, where the lack of investment in water or insufficient human capacity make it impossible for authorities to satisfy the demand for water. The report found that it would be possible to produce the food required in future, but that continuation of today's food production and environmental trends would lead to crises in many parts of the world. To avoid a global water crisis, farmers will have to strive to increase productivity to meet growing demands for food, while industry and cities find ways to use water more efficiently.

In some areas of the world, irrigation is necessary to grow any crop at all, in other areas it permits more profitable crops to be grown or enhances crop yield. Various irrigation methods involve different trade-offs between crop yield, water consumption and capital cost of equipment and structures. Irrigation methods such as furrow and overhead sprinkler irrigation are usually less expensive but are also typically less efficient, because much of the water evaporates, runs off or drains below the root zone. Other irrigation methods considered to be more efficient include drip or trickle irrigation, surge irrigation, and some types of sprinkler systems where the sprinklers are operated near ground level. These types of systems, while more expensive, usually offer greater potential to minimize runoff, drainage and evaporation. Any system that is improperly managed can be wasteful, all methods have the potential for high efficiencies under suitable conditions, appropriate irrigation timing and management. Some issues that are often insufficiently considered are salinization of groundwater and contaminant accumulation leading to water quality declines.

As global populations grow, and as demand for food increases in a world with a fixed water supply, there are efforts under way to learn how to produce more food with less water, through improvements in irrigation methods and technologies, agricultural water management, crop types, and water monitoring. Aquaculture is a small but growing agricultural use of water. Freshwater commercial fisheries may also be considered as agricultural uses of water, but have generally been assigned a lower priority than irrigation.

Industrial

It is estimated that 22% of worldwide water is used in industry. Major industrial users include hydroelectric dams, thermoelectric power plants, which use water for cooling, ore and oil refineries, which use water in chemical processes, and manufacturing plants, which use water as a solvent. Water withdrawal can be very high for certain industries, but consumption is generally much lower than that of agriculture.

A power plant in Poland

Water is used in renewable power generation. Hydroelectric power derives energy from the force of water flowing downhill, driving a turbine connected to a generator. This hydroelectricity is a low-cost, non-polluting, renewable energy source. Significantly, hydroelectric power can also be used for load following unlike most renewable energy sources which are intermittent. Ultimately, the energy in a hydroelectric powerplant is supplied by the sun. Heat from the sun evaporates water, which condenses as rain in higher altitudes and flows downhill. Pumped-storage hydroelectric plants also exist, which use grid electricity to pump water uphill when demand is low, and use the stored water to produce electricity when demand is high.

Hydroelectric power plants generally require the creation of a large artificial lake. Evaporation from this lake is higher than evaporation from a river due to the larger surface area exposed to the elements, resulting in much higher water consumption. The process of driving water through the turbine and tunnels or pipes also briefly removes this water from the natural environment, creating water withdrawal. The impact of this withdrawal on wildlife varies greatly depending on the design of the powerplant.

Pressurized water is used in water blasting and water jet cutters. Also, very high pressure water guns are used for precise cutting. It works very well, is relatively safe, and is not harmful to the environment. It is also used in the cooling of machinery to prevent overheating, or prevent saw blades from overheating. This is generally a very small source of water consumption relative to other uses.

Water is also used in many large scale industrial processes, such as thermoelectric power production, oil refining, fertilizer production and other chemical plant use, and natural gas extraction from shale rock. Discharge of untreated water from industrial uses is pollution. Pollution includes discharged solutes (chemical pollution) and increased water temperature (thermal pollution). Industry requires pure water for many applications and utilizes a variety of purification techniques both in water supply and discharge. Most of this pure water is generated on site, either from natural freshwater or from municipal grey water. Industrial consumption of water is generally much lower than withdrawal, due to laws requiring industrial grey water to be treated and returned to the environment. Thermoelectric powerplants using cooling towers have high consumption, nearly equal to their withdrawal, as most of the withdrawn water is evaporated as part of the cooling process. The withdrawal, however, is lower than in once-through cooling systems.

Domestic

Drinking water

It is estimated that 8% of worldwide water use is for domestic purposes. These include drinking water, bathing, cooking, toilet flushing, cleaning, laundry and gardening. Basic domestic water requirements have been estimated by Peter Gleick at around 50 liters per person per day, excluding water for gardens. Drinking water is water that is of sufficiently high quality so that it can be consumed or used without risk of immediate or long term harm. Such water is commonly called potable water. In most developed countries, the water supplied to domestic, commerce and industry is all of drinking water standard even though only a very small proportion is actually consumed or used in food preparation.

Recreation

Whitewater rapids

Sustainable management of water resources (including provision of safe and reliable supplies for drinking water and irrigation, adequate sanitation, protection of aquatic ecosystems, and flood protection) poses enormous challenges in many parts of the world.

Recreational water use is usually a very small but growing percentage of total water use. Recreational water use is mostly tied to reservoirs. If a reservoir is kept fuller than it would otherwise be for recreation, then the water retained could be categorized as recreational usage. Release of water from a few reservoirs is also timed to enhance whitewater boating, which also could be considered a recreational usage. Other examples are anglers, water skiers, nature enthusiasts and swimmers.

Recreational usage is usually non-consumptive. Golf courses are often targeted as using excessive amounts of water, especially in drier regions. It is, however, unclear whether recreational irrigation (which would include private gardens) has a noticeable effect on water resources. This is largely due to the unavailability of reliable data. Additionally, many golf courses utilize either primarily or exclusively treated effluent water, which has little impact on potable water availability.

Some governments, including the Californian Government, have labelled golf course usage as agricultural in order to deflect environmentalists' charges of wasting water. However, using the above figures as a basis, the actual statistical effect of this reassignment is close to zero. In Arizona, an organized lobby has been established in the form of the Golf Industry Association, a group focused on educating the public on how golf impacts the environment.

Recreational usage may reduce the availability of water for other users at specific times and places. For example, water retained in a reservoir to allow boating in the late summer is not available to farmers during the spring planting season. Water released for whitewater rafting may not be available for hydroelectric generation during the time of peak electrical demand.

Environmental

Explicit environment water use is also a very small but growing percentage of total water use. Environmental water may include water stored in impoundments and released for environmental purposes (held environmental water), but more often is water retained in waterways through regulatory limits of abstraction. Environmental water usage includes watering of natural or artificial wetlands, artificial lakes intended to create wildlife habitat, fish ladders, and water releases from reservoirs timed to help fish spawn, or to restore more natural flow regimes

Like recreational usage, environmental usage is non-consumptive but may reduce the availability of water for other users at specific times and places. For example, water release from a reservoir to help fish spawn may not be available to farms upstream, and water retained in a river to maintain waterway health would not be available to water abstractors downstream.

Water Stress

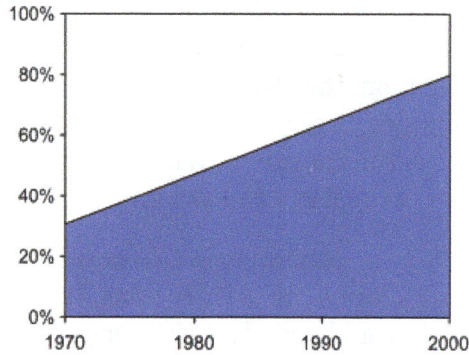

Estimate of the share of people in developing countries with access to drinking water 1970–2000

The concept of water stress is relatively simple: According to the World Business Council for Sustainable Development, it applies to situations where there is not enough water for all uses, whether agricultural, industrial or domestic. Defining thresholds for stress in terms of available water per capita is more complex, however, entailing assumptions about water use and its efficiency. Nevertheless, it has been proposed that when annual per capita renewable freshwater availability is less than 1,700 cubic meters, countries begin to experience periodic or regular water stress. Below 1,000 cubic meters, water scarcity begins to hamper economic development and human health and well-being.

Population Growth

In 2000, the world population was 6.2 billion. The UN estimates that by 2050 there will be an additional 3.5 billion people with most of the growth in developing countries that already suffer water stress. Thus, water demand will increase unless there are corresponding increases in water conservation and recycling of this vital resource. In building on the data presented here by the UN, the World Bank goes on to explain that access to water for producing food will be one of the main challenges in the decades to come. Access to water will need to be balanced with the importance of managing water itself in a sustainable way while taking into account the impact of climate change, and other environmental and social variables.

Expansion of Business Activity

Business activity ranging from industrialization to services such as tourism and entertainment continues to expand rapidly. This expansion requires increased water services including both supply and sanitation, which can lead to more pressure on water resources and natural ecosystem.

Rapid Urbanization

The trend towards urbanization is accelerating. Small private wells and septic tanks that work well in low-density communities are not feasible within high-density urban areas. Urbanization requires significant investment in water infrastructure in order to deliver water to individuals and to process the concentrations of wastewater – both from individuals and from business. These polluted and contaminated waters must be treated or they pose unacceptable public health risks.

In 60% of European cities with more than 100,000 people, groundwater is being used at a faster rate than it can be replenished. Even if some water remains available, it costs increasingly more to capture it.

Climate Change

Climate change could have significant impacts on water resources around the world because of the close connections between the climate and hydrological cycle. Rising temperatures will increase evaporation and lead to increases in precipitation, though there will be regional variations in rainfall. Both droughts and floods may become more frequent in different regions at different times, and dramatic changes in snowfall and snow melt are expected in mountainous areas. Higher temperatures will also affect water quality in ways that are not well understood. Possible impacts include increased eutrophication. Climate change could also mean an increase in demand for farm irrigation, garden sprinklers, and perhaps even swimming pools. There is now ample evidence that increased hydrologic variability and change in climate has and will continue have a profound impact on the water sector through the hydrologic cycle, water availability, water demand, and water allocation at the global, regional, basin, and local levels.

Depletion of Aquifers

Due to the expanding human population, competition for water is growing such that many of the world's major aquifers are becoming depleted. This is due both for direct human consumption as well as agricultural irrigation by groundwater. Millions of pumps of all sizes are currently extracting groundwater throughout the world. Irrigation in dry areas such as northern China, Nepal and India is supplied by groundwater, and is being extracted at an unsustainable rate. Cities that have experienced aquifer drops between 10 and 50 meters include Mexico City, Bangkok, Beijing, Madras and Shanghai.

Pollution and Water Protection

Water pollution is one of the main concerns of the world today. The governments of numerous countries have striven to find solutions to reduce this problem. Many pollutants threaten water supplies, but the most widespread, especially in developing countries, is the discharge of raw sewage into natural waters; this method of sewage disposal is the most common method in underdeveloped countries, but also is prevalent in quasi-developed countries such as China, India, Nepal and Iran. Sewage, sludge, garbage, and even toxic pollutants are all dumped into the water. Even if sewage is treated, problems still arise. Treated sewage forms sludge, which may be placed in landfills, spread out on land, incinerated or dumped at sea. In addition to sewage, nonpoint source pollution such as agricultural runoff is a significant source of pollution in some parts of the world, along with urban stormwater runoff and chemical wastes dumped by industries and governments.

Polluted water

Water and Conflicts

Competition for water has widely increased, and it has become more difficult to conciliate the necessities for water supply for human consumption, food production, ecosystems and other uses. Water administration is frequently involved in contradictory and complex problems. Approximately 10% of the worldwide annual runoff is used for human necessities. Several areas of the world are flooded, while others have such low precipitations that human life is almost impossible. As population and development increase, raising water demand, the possibility of problems inside a certain country or region increases, as it happens with others outside the region.

Over the past 25 years, politicians, academics and journalists have frequently predicted that disputes over water would be a source of future wars. Commonly cited quotes include: that of former Egyptian Foreign Minister and former Secretary-General of the United Nations Boutrous Ghali, who forecast, "The next war in the Middle East will be fought over water, not politics"; his successor at the UN, Kofi Annan, who in 2001 said, "Fierce competition for fresh water may well become a source of conflict and wars in the future," and the former Vice President of the World Bank, Ismail Serageldin, who said the wars of the next century will be over water unless significant changes in governance occurred. The water wars hypothesis had its roots in earlier research carried out on a small number of transboundary rivers such as the Indus, Jordan and Nile. These particular rivers became the focus because they had experienced water-related disputes. Specific events cited as evidence include Israel's bombing of Syria's attempts

to divert the Jordan's headwaters, and military threats by Egypt against any country building dams in the upstream waters of the Nile. However, while some links made between conflict and water were valid, they did not necessarily represent the norm.

The only known example of an actual inter-state conflict over water took place between 2500 and 2350 BC between the Sumerian states of Lagash and Umma. Water stress has most often led to conflicts at local and regional levels. Tensions arise most often within national borders, in the downstream areas of distressed river basins. Areas such as the lower regions of China's Yellow River or the Chao Phraya River in Thailand, for example, have already been experiencing water stress for several years. Water stress can also exacerbate conflicts and political tensions which are not directly caused by water. Gradual reductions over time in the quality and/or quantity of fresh water can add to the instability of a region by depleting the health of a population, obstructing economic development, and exacerbating larger conflicts.

Shared Water Resources can Promote Collaboration

Water resources that span international boundaries are more likely to be a source of collaboration and cooperation than war. Scientists working at the International Water Management Institute have been investigating the evidence behind water war predictions. Their findings show that, while it is true there has been conflict related to water in a handful of international basins, in the rest of the world's approximately 300 shared basins the record has been largely positive. This is exemplified by the hundreds of treaties in place guiding equitable water use between nations sharing water resources. The institutions created by these agreements can, in fact, be one of the most important factors in ensuring cooperation rather than conflict.

The International Union for the Conservation of Nature (IUCN) published the book *Share: Managing water across boundaries*. One chapter covers the functions of trans-boundary institutions and how they can be designed to promote cooperation, overcome initial disputes and find ways of coping with the uncertainty created by climate change. It also covers how the effectiveness of such institutions can be monitored.

Water Shortages

In 2025, water shortages will be more prevalent among poorer countries where resources are limited and population growth is rapid, such as the Middle East, Africa, and parts of Asia. By 2025, large urban and peri-urban areas will require new infrastructure to provide safe water and adequate sanitation. This suggests growing conflicts with agricultural water users, who currently consume the majority of the water used by humans.

Generally speaking the more developed countries of North America, Europe and Russia will not see a serious threat to water supply by the year 2025, not only because of their relative wealth, but more importantly their populations will be better aligned with available water resources. North Africa, the Middle East, South Africa and northern China will face very severe water shortages due to physical scarcity and a condition of overpopulation relative to their carrying capacity with respect to water supply. Most of South America, Sub-Saharan Africa, Southern China and India will face water supply shortages by 2025; for these latter regions the causes of scarcity will be economic constraints to developing safe drinking water, as well as excessive population growth.

Economic Considerations

Water supply and sanitation require a huge amount of capital investment in infrastructure such as pipe networks, pumping stations and water treatment works. It is estimated that Organisation for Economic Co-operation and Development (OECD) nations need to invest at least USD 200 billion per year to replace aging water infrastructure to guarantee supply, reduce leakage rates and protect water quality.

International attention has focused upon the needs of the developing countries. To meet the Millennium Development Goals targets of halving the proportion of the population lacking access to safe drinking water and basic sanitation by 2015, current annual investment on the order of USD 10 to USD 15 billion would need to be roughly doubled. This does not include investments required for the maintenance of existing infrastructure.

Once infrastructure is in place, operating water supply and sanitation systems entails significant ongoing costs to cover personnel, energy, chemicals, maintenance and other expenses. The sources of money to meet these capital and operational costs are essentially either user fees, public funds or some combination of the two. An increasing dimension to consider is the flexibility of the water supply system.

Water Resource Management

Water resource management is the activity of planning, developing, distributing and managing the optimum use of water resources. It is a sub-set of water cycle management. Ideally, water resource management planning has regard to all the competing demands for water and seeks to allocate water on an equitable basis to satisfy all uses and demands. As with other resource management, this is rarely possible in practice.

Overview

Water is an essential resource for all life on the planet. Of the water resources on Earth, only three percent of it is fresh and two-thirds of the freshwater is locked up in ice caps and glaciers. Of the remaining one percent, a fifth is in remote, inaccessible areas and much seasonal rainfall in monsoonal deluges and floods cannot easily be used. As time advances, water is becoming scarcer and having access to clean, safe, drinking water is limited among countries. At present only about 0.08 percent of all the world's fresh water is exploited by mankind in ever increasing demand for sanitation, drinking, manufacturing, leisure and agriculture. Due to the small percentage of water remaining, optimizing the fresh water we have left from natural resources has been a continuous difficulty in several locations worldwide.

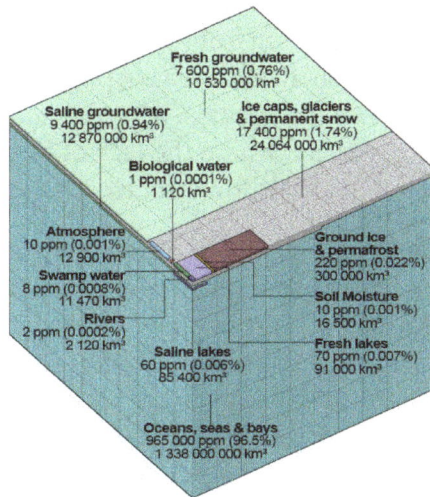

Fresh groundwater
7 600 ppm (0.76%)
10 530 000 km³

Saline groundwater
9 400 ppm (0.94%)
12 870 000 km³

Ice caps, glaciers
& permanent snow
17 400 ppm (1.74%)
24 064 000 km³

Biological water
1 ppm (0.0001%)
1 120 km³

Atmosphere
10 ppm (0.001%)
12 900 km³

Ground ice
& permafrost
220 ppm (0.022%)
300 000 km³

Swamp water
8 ppm (0.0008%)
11 470 km³

Soil Moisture
10 ppm (0.001%)
16 500 km³

Rivers
2 ppm (0.0002%)
2 120 km³

Saline lakes
60 ppm (0.006%)
85 400 km³

Fresh lakes
70 ppm (0.007%)
91 000 km³

Oceans, seas & bays
965 000 ppm (96.5%)
1 338 000 000 km³

Visualisation of the distribution (by volume) of water on Earth. Each tiny cube (such as the one representing biological water) corresponds to approximately 1,000 cubic kilometres (240 cu mi) of water, with a mass of approximately 1 trillion tonnes (2000 times that of the Great Pyramid of Giza or 5 times that of Lake Kariba, arguably the heaviest man-made object). The entire block comprises 1 million tiny cubes.

Much effort in water resource management is directed at optimizing the use of water and in minimizing the environmental impact of water use on the natural environment. The observation of water as an integral part of the ecosystem is based on integrated water resource management, where the quantity and quality of the ecosystem help to determine the nature of the natural resources.

Successful management of any resources requires accurate knowledge of the resource available, the uses to which it may be put, the competing demands for the resource, measures to and processes to evaluate the significance and worth of competing demands and mechanisms to translate policy decisions into actions on the ground.

For water as a resource, this is particularly difficult since sources of water can cross many national boundaries and the uses of water include many that are difficult to assign financial value to and may also be difficult to manage in conventional terms. Examples include rare species or ecosystems or the very long term value of ancient groundwater reserves.

Agriculture

Agriculture is the largest user of the world's freshwater resources, consuming 70 percent. As the world population rises it consumes more food (currently exceeding 6%, it is expected to reach 9% by 2050), the industries and urban developments expand, and the emerging biofuel crops trade also demands a share of freshwater resources, water scarcity is becoming an important issue. An assessment of water resource management in agriculture was conducted in 2007 by the International Water Management Institute in Sri Lanka to see if the world had sufficient water to provide food for its growing population or not . It assessed the current availability of water for agriculture on a global scale and mapped out locations suffering from water scarcity. It found that a fifth of the world's people, more than 1.2 billion, live in areas of physical water scarcity, where there is not enough water to meet all their demands. A further 1.6 billion people live in areas experiencing economic water scarcity, where the lack of investment in water or insufficient human capacity make it impossible for authorities to satisfy the demand for water.

The report found that it would be possible to produce the food required in future, but that continuation of today's food production and environmental trends would lead to crises in many parts of the world. Regarding food production, the World Bank targets agricultural food production and water resource management as an increasingly global issue that is fostering an important and growing debate. The authors of the book *Out of Water: From abundance to Scarcity and How to Solve the World's Water Problems*, which laid down a six-point plan for solving the world's water problems. These are: 1) Improve data related to water; 2) Treasure the environment; 3) Reform water governance; 4) Revitalize agricultural water use; 5) Manage urban and industrial demand; and 6) Empower the poor and women in water resource management. To avoid a global water crisis, farmers will have to strive to increase productivity to meet growing demands for food, while industry and cities find ways to use water more efficiently.

Managing Water in Urban Settings

As the carrying capacity of the Earth increases greatly due to technological advances, urbanization in modern times occurs because of economic opportunity. This rapid urbanization happens worldwide but mostly in new rising economies and developing countries. Cities in Africa and Asia are growing fastest with 28 out of 39 megacities (a city or urban area with more than 10 million inhabitants) worldwide in these developing nations. The number of megacities will continue to rise reaching approximately 50 in 2025. With developing economies water scarcity is a very common and very prevalent issue. Global freshwater resources dwindle in the eastern hemisphere either than at the poles, and with the majority of urban development millions live with insufficient fresh water. This is caused by polluted freshwater resources, overexploited groundwater resources, insufficient harvesting capacities in the surrounding rural areas, poorly constructed and maintained water supply systems, high amount of informal water use and insufficient technical and water management capacities.

In the areas surrounding urban centres, agriculture must compete with industry and municipal users for safe water supplies, while traditional water sources are becoming polluted with urban runoff. As cities offer the best opportunities for selling produce, farmers often have no alternative to using polluted water to irrigate their crops.

Depending on how developed a city's wastewater treatment is, there can be significant health hazards related to the use of this water. Wastewater from cities can contain a mixture of pollutants. There is usually wastewater from kitchens and toilets along with rainwater runoff. This means that the water usually contains excessive levels of nutrients and salts, as well as a wide range of pathogens. Heavy metals may also be present, along with traces of antibiotics and endocrine disruptors, such as oestrogens.

Developing world countries tend to have the lowest levels of wastewater treatment. Often, the water that farmers use for irrigating crops is contaminated with pathogens from sewage. The pathogens of most concern are bacteria, viruses and parasitic worms, which directly affect farmers' health and indirectly affect consumers if they eat the contaminated crops. Common illnesses include diarrhoea, which kills 1.1 million people annually and is the second most common cause of infant deaths. Many cholera outbreaks are also related to the reuse of poorly treated wastewater. Actions that reduce or remove contamination, therefore, have the potential to save a large number of lives and improve livelihoods. Scientists have been working to find ways to reduce contamination of food using a method called the 'multiple-barrier approach'.

This involves analysing the food production process from growing crops to selling them in markets and eating them, then considering where it might be possible to create a barrier against contamination. Barriers include: introducing safer irrigation practices; promoting on-farm wastewater treatment; taking actions that cause pathogens to die off; and effectively washing crops after harvest in markets and restaurants.

Urban Decision Support System (UDSS)

Urban Decision Support System (UDSS) – is a wireless device with a mobile app that uses sensors attached to water appliances in urban residences to collect data about water usage and is an example of data-driven urban water management. The system was developed with a European Commission investment of 2.46 Million Euros to improve the water consumption behaviour of households. Information about every mechanism – dishwashers, showers, washing machines, taps – is wirelessly recorded and sent to the UDSS App on the user's mobile device. The UDSS is then able to analyse and show homeowners which of their appliances are using the most water, and which behaviour or habits of the households are not encouraged in order to reduce the water usage, rather than simply giving a total usage figure for the whole property, which will allow people to manage their consumption more economically. The UDSS is based on university research in the field of Management Science, at Loughborough University School of Business and Economics, particularly Decision Support System in household water benchmarking, lead by Dr Lili Yang, (Reader)

Future of Water Resources

One of the biggest concerns for our water-based resources in the future is the sustainability of the current and even future water resource allocation. As water becomes more scarce, the importance of how it is managed grows vastly. Finding a balance between what is needed by humans and what is needed in the environment is an important step in the sustainability of water resources. Attempts to create sustainable freshwater systems have been seen on a national level in countries such as Australia, and such commitment to the environment could set a model for the rest of the world.

The field of water resources management will have to continue to adapt to the current and future issues facing the allocation of water. With the growing uncertainties of global climate change and the long term impacts of management actions,the decision-making will be even more difficult. It is likely that ongoing climate change will lead to situations that have not been encountered. As a result, alternative management strategies are sought for in order to avoid setbacks in the allocation of water resources.

References

- The World Bank, 2009 "Water and Climate Change: Understanding the Risks and Making Climate-Smart Investment Decisions". Retrieved 2011-10-24

- Klimaszyk Piotr, Rzymski Piotr "Surface Runoff as a Factor Determining Trophic State of Midforest Lake" Polish Journal of Environmental Studies, 2011, 20(5), 1203-1210

- H. Edward Reiley; Carroll L. Shry (2002). Introductory horticulture. Cengage Learning. p. 40. ISBN 978-0-7668-1567-4

- FMI (2007). "Fog And Stratus - Meteorological Physical Background". Zentralanstalt für Meteorologie und Geodynamik. Retrieved 2009-02-07

- J. S. oguntoyinbo and F. 0. Akintola (1983). "Rainstorm characteristics affecting water availability for agriculture" (PDF). IAHS Publication Number 140. Retrieved 2008-12-27

- Allen, R.G.; Pereira, L.S.; Raes, D.; Smith, M. (1998). Crop Evapotranspiration: Guidelines for Computing Crop Water Requirements. FAO Irrigation and drainage paper 56. Rome, Italy: Food and Agriculture Organization of the United Nations. ISBN 92-5-104219-5. Retrieved 2011-06-08

- Zhang, S.X.; V. Babovic (2012). "A real options approach to the design and architecture of water supply systems using innovative water technologies under uncertainty" (PDF). Journal of Hydroinformatics

- Jennifer E. Lawson (2001). Hands-on Science: Light, Physical Science (matter) - Chapter 5: The Colors of Light. Portage & Main Press. p. 39. ISBN 978-1-894110-63-1. Retrieved 2009-06-28

- Norman W. Junker (2008). "An ingredients based methodology for forecasting precipitation associated with MCS's". Hydrometeorological Prediction Center. Retrieved 2009-02-07

- S. I. Efe; F. E. Ogban; M. J. Horsfall; E. E. Akporhonor (2005). "Seasonal Variations of Physico-chemical Characteristics in Water Resources Quality in Western Niger Delta Region, Nigeria" (PDF). Journal of Applied Scientific Environmental Management. 9 (1): 191–195. ISSN 1119-8362. Retrieved 2008-12-27

- Nouri, Hamideh; Beecham, Simon; Anderson, Sharoyn; Hassanli, Morad; Kazemi, Fatemeh (13 May 2014). "Remote sensing techniques for predicting evapotranspiration from mixed vegetated surfaces". Urban Water J. doi:10.1080/1573062X.2014.900092

- Toby N. Carlson (1991). Mid-latitude Weather Systems. Routledge. p. 216. ISBN 978-0-04-551115-0. Retrieved 2009-02-07

- National Severe Storms Laboratory (2007-04-23). "Aggregate hailstone". National Oceanic and Atmospheric Administration. Retrieved 2009-07-15

- Yuh-Lang Lin (2007). Mesoscale Dynamics. Cambridge University Press. p. 405. ISBN 978-0-521-80875-0. Retrieved 2009-07-07

- Climate Prediction Center (2005). "2005 Tropical Eastern North Pacific Hurricane Outlook". National Oceanic and Atmospheric Administration. Retrieved 2006-05-02

- C. D. Haynes; M. G. Ridpath; M. A. J. Williams (1991). Monsoonal Australia. Taylor & Francis. p. 90. ISBN 978-90-6191-638-3. Retrieved 2008-12-27

- Michael Ritter (2008-12-24). "Subarctic Climate". University of Wisconsin–Stevens Point. Archived from the original on 2008-05-25. Retrieved 2008-04-16

- A. Roberto Frisancho (1993). Human Adaptation and Accommodation. University of Michigan Press, pp. 388. ISBN 978-0-472-09511-7. Retrieved on 2008-12-27

- Robert Burns (2007-06-06). "Texas Crop and Weather". Texas A&M University. Archived from the original on 2010-06-20. Retrieved 2010-01-15

2

Significance of Modelling in Water Resources

Water resources modeling is extremely significant and useful when preservation of water and its quality is concerned. A few models such as river basin planning and management, water distribution system, water quality modeling, etc. are discussed in this chapter. In river basin, surface water and groundwater are submerged together, due to which its planning and management gains importance. The major components of modeling in water resources are discussed in this section.

River Basin

A river basin can be defined as the geographical area demarcated by the topographic limits of the system of waters. There are strong interactions between land and water resources, or between surface and groundwater in the basin. Hence, river basins are important elements in water resource development and planning. It possesses freshwater and also shapes up the landscapes. Most of the objectives discussed so far such as irrigation, hydropower generation, recreation, navigation etc take place in the basin itself. The upstream basin characteristic influences the availability of water for the above said purposes. River basins are also the most productive ecosystems. In this lesson, we will discuss more about the river basin management.

River Basin Management (RBM)

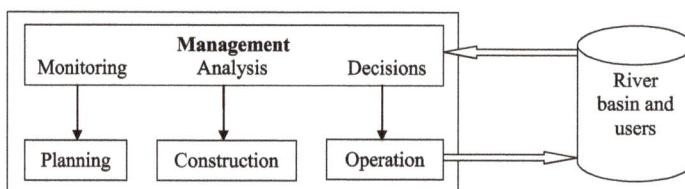

Activities in RBM

RBM is essentially the management of water resources in the basin. The purpose of RBM is to ensure the use of water and other resources in the basin in a sustainable manner. Most of the basins have multi-objectives and limited resources. Hence, it is necessary to assign priorities to different needs. The need for RBM arises from the non-coordinated usage or even overexploitation of resources. RBM can be divided into six activities: planning, construction, operation, monitoring, analysis and decision making as shown in the figure above.

The basin and its users are directly affected by operation and management activities only. The operation activity as such has a direct influence on the basin. This includes regulation of facilities and application of economic, legal and policy instruments. This is conducted through analysis and decision support systems (DSS). RBM may change the basin characteristics by the construction of storage or diversion structures; by implementing allocation rules, water rights and permits; by imposing taxes and subsidies to control water usage. RBM may also contain a number of related activities like public participation, international participation or cooperation between related organizations. The most important structural regulation in RBM is the reservoir operation. A very sensitive operational issue in RBM is charging the users for water usage. This is an effective means to minimize wastage and also it will finance development activities in the basin. However, the charges should be low enough for the poor to afford.

River Basin Planning

Planning is an important and inevitable part in the utilization of water resources and in the operation of projects. It helps to assess the present situation and achieve the desired situation by filling the gaps. It offers a framework by setting priorities and also results in more public support. This brings together managers of different river basins together, thus resulting in a common goal. Planning has four functions:

(a) To assess the current situation, formulate and set goals and targets, orient the operation and management.

(b) To provide framework for public participation and feedback

(c) To increase legitimacy and mobilize public acceptance

(d) To facilitate interaction among the concerned organizations and stakeholders

The important activities of planning of river basins are:

(a) Identification of need, scope and geographical coverage area.

(b) Analysis of institutional framework, identification of decisions, and the formulation of bodies.

(c) Identification of stakeholders, their preferences and expectations

(d) Preparation of blueprint of scope, identify different phases and groups involved, prepare a flow chart of activities

(e) Formulation of plans and its approval

(f) Implementation of plan

The types of plan for a particular river basin depend on different factors such as policy issues, location, available funds etc. These factors differ from country to country and basin to basin. RBM should also consider the interrelations within water systems (i.e., surface, and subsurface; quantity and quality), the interrelations between climate, land and water and also interrelations between complete river basins and their socio-economic environment. The planning depends on the number of functions to be performed. If the main goal is drinking water supply for an urban city, then there is no need to consider strategic planning which may require an overall description of the basin. If too many plans are to be executed simultaneously, then the availability of resources and also coordination can become a problem.

Integrated Water Resources Management (IWRM)

A river basin can be divided into three components: source components (rivers, reservoirs, aquifers and canals), demand components (off-stream such as irrigation fields, industrial plants and cities and in-stream such as hydropower, recreation and environment) and intermediate components (treatment plants, reuse and recycling facilities).

According to the Technical Advisory Committee (TAC) Global partnership 2000, IWRM is a process which promotes the coordinated development and management of water, land and related resources, in order to maximize the resultant economic and social welfare in an equitable manner without compromising the sustainability of vital ecosystems. IWRM is both a goal and also a process of balancing and making trade-offs between different goals in an informed way. IWRM integrates three systems: natural, socio-economic and institutional. IWRM manages the conflict that may occur between the physical and social linkages of a water system.

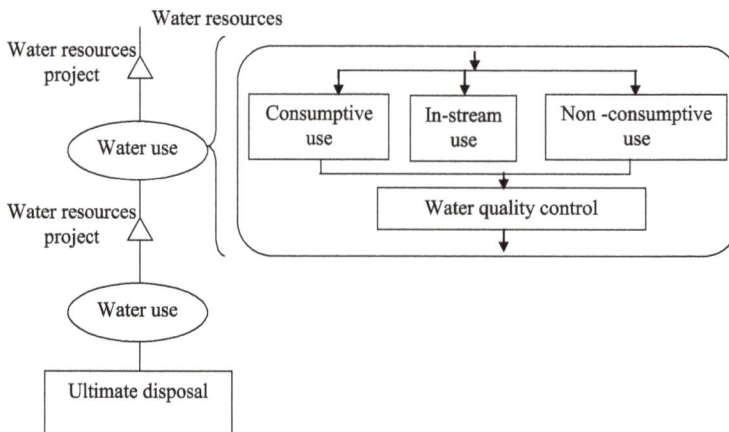

Steps in water resources management

The steps for water management in a basin are shown in the above figure. Along with managing various projects in a basin, IWRM also manages utilization of water for consumptive uses, non-consumptive uses and in-stream uses.

Models for IWRM

First generation models concentrated in hydraulic and hydrologic aspects such as flood routing, reservoir operation etc. Models for sediment transportation and water quality simulation were developed side by side. These include Streamflow Synthesis And Reservoir Regulation (SSARR) from United States Army Corps of Engineers (USACE) and SIMLYD – II from Texas water development board. HEC-5 model is also widely used to simulate operation of reservoir systems.

Second generation models are able to consider both hydrologic and water quality aspects. These are able to perform interactive analysis and display of results. The Interactive River – Aquifer Simulation (IRAS) (Loucks et al., 1994) extensively uses graphics in system simulation. The European Hydrological System (SHE) is a distributed and physically modeled system and describes the major land flow processes of the hydrologic cycle. MIKE SHE is a advanced version of SHE with add-ons for water quality, soil erosion, irrigation etc. Water quality simulation models are a standard feature of river basin models. A widely applied one is the Enhanced Stream Water Quality Model (QUAL2E) which simulates temperature, dissolved oxygen, biochemical oxygen demands etc. Another one is the Water Quality for River Reservoir Systems (WQRSS) by Hydrologic Engineering Center.

The third generation models are interactive models with graphical user interfaces, GIS inputs and screen display of results. The Tennessee river valley authority Environment and River Resource Aid (TERRA) model is a reservoir and power generation management tool linked to a local area network for real time functioning. However, it is designed to function on Tennessee river valley.

River Basin Simulation Model (RIBASIM) simulates river basins for various hydrological conditions. This model links hydrological inputs at various locations with specific water uses in the basin. It also evaluates alternatives of infrastructure, operational and demand management through a decision support system.

The MIKE BASIN represents rivers and tributaries as network with branches and nodes. This has a graphical interface which is linked to ArcView GIS. This gives information on individual reservoir outputs and irrigation scheme with frequency and magnitude of water shortages.

MIKE SHE is an advanced integrated hydrological modeling system which simulates water flow in the entire land based phase of the hydrological cycle from rainfall to river flow, via various flow processes such as, overland flow, infiltration into soils, evapotranspiration from vegetation, and groundwater flow. The hydrologic processes simulated by MIKE SHE are shown in the figure.

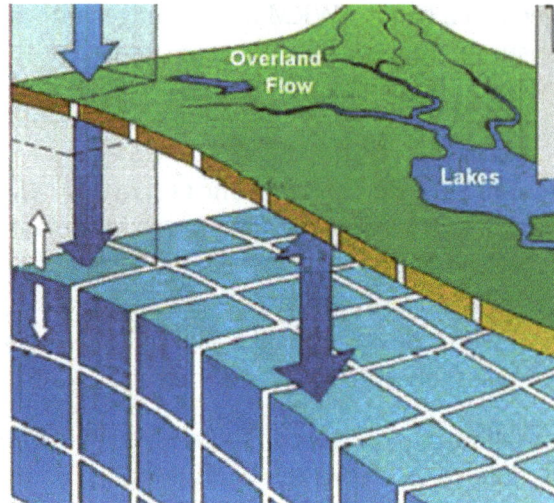

MIKE SHE is linked with ArcView GIS for pre- and post-processing of data. The main characteristics are

   (i) Integrated: Fully dynamic exchange of water between all major hydrological components is included, e.g. surface water, soil water and groundwater

  (ii) Physically based: Solves basic equations governing the major flow processes within the study area

 (iii) Fully distributed: The spatial and temporal variation of meteorological, hydrological, geological and hydrogeological data across the model area is described in gridded form for the input as well as the output from the model

  (iv) Modular: It has a modular structure, which allows user to focus only on the processes, which are important for the study.

MIKE SHE has been used for the analysis, planning and management of a wide range of water resources problems such as:

  • River basin management and planning

  • Water supply design, management and optimization

  • Irrigation and drainage

  • Soil and water management

  • Conjunctive use of groundwater and surface water

  • Groundwater management

  • Contamination from waste disposal

- Floodplain studies

- Impact of land use and climate change etc.

MIKE BASIN is a modeling tool for integrated river basin planning and management developed by Danish Hydraulic Institute (DHI) in Denmark. It addresses water allocation, conjunctive use, reservoir operation, and water quality issues. It couples ArcGIS with hydrologic modeling to provide basin-scale solutions. It provides a mathematical representation of the river basin encompassing the main rivers and their tributaries, the hydrology of the basin in space and time, existing as well as potential major schemes and their various demands of water.

River systems are represented by a network consisting of branches and nodes. Branches represent individual stream sections while the nodes represent confluences, locations where certain water activities may occur, or important locations where model results are required. This model gives emphasis on both simulation and visualization in both space and time. Typical areas of application include water availability analysis, conjunctive use of surface and groundwater, infrastructure planning, assessing irrigation potential and reservoir performance, estimating water supply capacity, determining waste water treatment requirements. The Graphical User Interfaces (GUI) for different applications are shown in figures below.

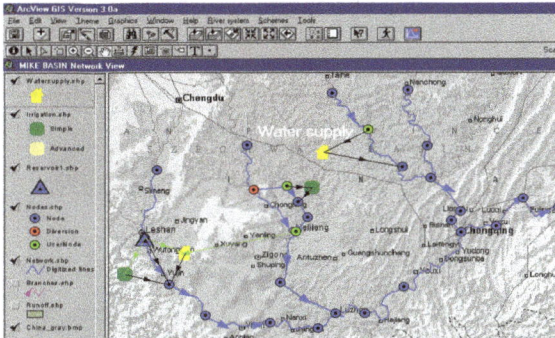

Representation of river network in a basin

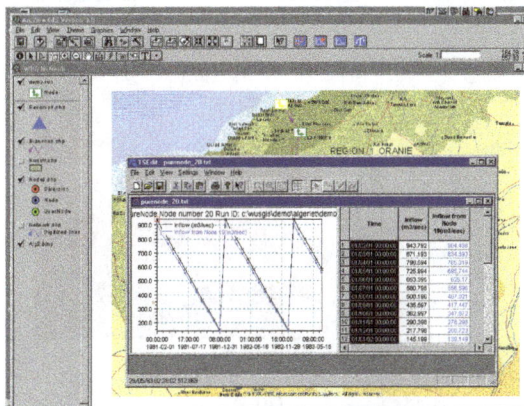

Illustration of results – Flow at a node

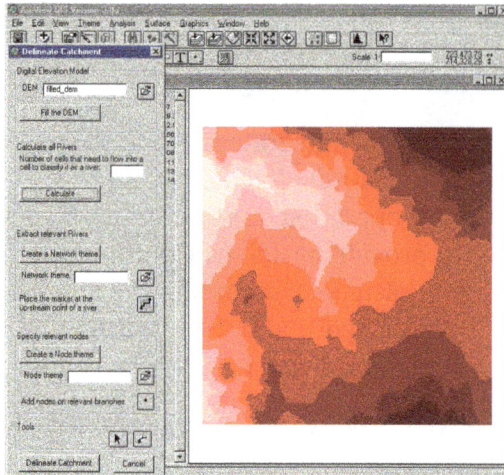

(a)

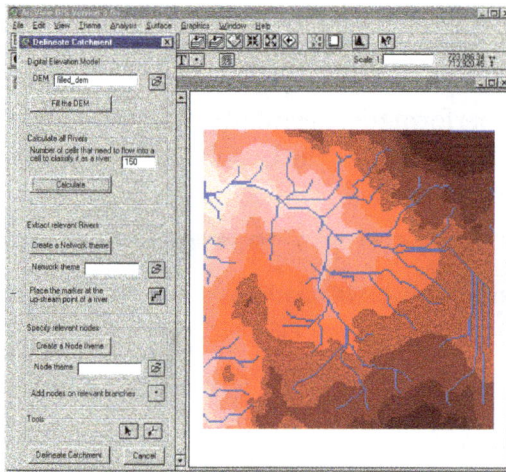

(b)

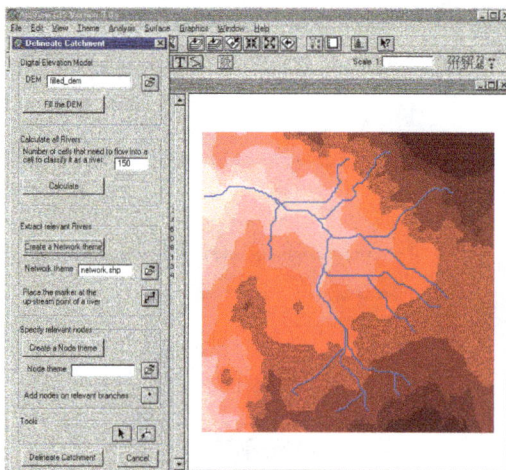

(c)

Delineation of catchment and river networks from digital elevation models (DEMs)

Water Supply

The main purpose of water distribution network is to supply water to the users according to their demand with adequate pressure. Water distribution systems are composed of three major components: pumping stations, storage tanks and distribution piping. These systems are designed according to the loading conditions i.e., pressure and demand at nodal points. The loading conditions may include fire demands, peak daily demands or critical demands when the pipes are broken. A reliable design should consider all the loading conditions including the critical conditions. In this lesson we will discuss the simulation and optimization models for the design and analysis of water distribution networks.

Components of Water Distribution Systems

Various components of water distribution systems are:

(i) Pipes: These are the principal elements in the system. The flow or velocity is usually described using Hazen – Williams equation

$$V = 1.318 C_{HW} R^{0.63} S_f^{0.54}$$

where V is the average flow velocity. C_{HW} is the Hazen – Williams roughness coefficient, R is the hydraulic radius and Sf is the slope.

In terms of headloss h_L, the above equation can be expressed as,

$$h_L = \frac{KLQ^{1.852}}{C_{HW}^{1.852} D^{4.87}} = K_P Q^{1.852}$$

where L is the length of the pipe, D is the diameter and Q is the flow rate. Headloss can also be determined using Darcy – Weisbach equation as

$$h_L = f \frac{L}{D} \frac{V^2}{2g} = \frac{8fL}{\pi^2 g D^2} Q^2 K_P Q^2$$

where f is the friction factor (determined from Moody's diagram) and g is the acceleration due to gravity.

(ii) Node: Junction nodes are connections of pipes to transfer the water. The diameter of pipe is changed at these nodes. Fixed grade nodes is where pressure and elevation are fixed i.e., reservoirs, tanks etc.

(iii) Valves: These are used to vary the head loss or to control the flow.

(iv) Tanks: It stores water and acts as a buffer by storing water at low demands and releasing at high demands.

(v) Pumps: Used to increase the energy

Simulation of Network

The flow distribution through a network should satisfy the conservation of mass and conservation of energy. Consider the network structure in the given figure with 6 pipes and 5 nodes.

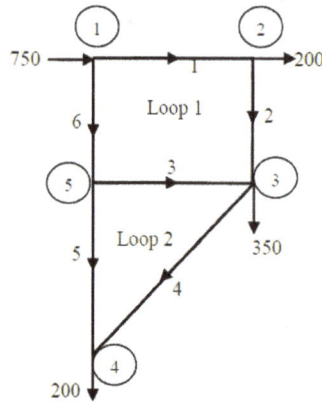

Conservation of mass: Flow at each junction nodes must be conserved

$$\sum Q_{in} - \sum Q_{out} = Q_{ext}$$

where Q_{in} and Q_{out} are the flows in and out of the node respectively and Q_{ext} is the external supply or demand.

Conservation of energy: For each loop, energy must be conserved i.e., sum of head losses should be zero.

$$\sum h_{L_{i,j}} - \sum H_{pump} = 0$$

where $h_{L_{i,j}}$ is the head loss in the pipe connecting nodes i and j and H_{pump} is the energy added by the pump (if any). $h_{L_{i,j}}$ can be determined using equations above.

Energy must be conserved between the fixed grade nodes which are points of known head (elevation plus pressure head).

$$\Delta E_F = \sum h_{L_{i,j}} - \sum H_{pump}$$

If the number of pipes in the network is N_L, number of junction nodes is N_J and number of fixed grade nodes is N_F, then total number of equations will be $N_L + N_J + (N_F - 1)$.

The set of equations obtained can be solved by any iterative techniques like Hardy-Cross method, linear theory method and Newton - Raphson method.

Hardy-Cross Method

In this method, the loop equation in terms of flow is used. The loop equations are transformed into so called ΔQ equations in the form

$$\sum_{i,j} K_{p,i,j} Q_{i,j} + \Delta Q_{i,j} = 0$$

Here head loss is determined from equation earlier.

Equation above is rewritten to account the direction of flow as

$$\sum h_{L_{i,j}} = \sum_{i,j} K_{p,i,j} Q_{i,j} + \Delta Q_{i,j} \; sign \, Q_{i,j} + \Delta Q_{i,j} = 0$$

This equation can be finally expressed as

$$\Delta Q_p = -\frac{\displaystyle\sum_{i,j} K_{p,i,j} Q_{i,j}^n}{\displaystyle\sum_{i,j} \left| n K_{p,i,j} Q_{i,j}^{n-1} \right|}$$

First a flow distribution is assumed across the network. Then the correction ΔQ as given in eqn. above is applied in a particular loop p. The numerator in eqn. above is the algebraic sum of headloss in loop p taking care of the sign of the flow. If clockwise flows are taken positive, then the corresponding headlosses are positive. The same is applicable while applying correction also i.e., ΔQ_p is added to flows in the clockwise direction and subtracted from flows in counterclockwise direction.

Example:

Consider the pipe network shown below. The friction factor = 0.2. Determine the flow rate in each pipe.

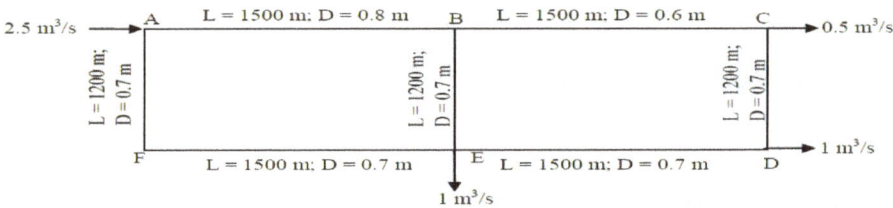

Solution:

Step 1: Determine the K values in equation.

$$h_L = \frac{8fL}{\pi^2 gD^2}Q^2 = KQ^2 \quad \text{where } K = \frac{8fL}{\pi^2 gD^2}$$

$$K_{AB} = 3.88; \quad K_{BC} = 6.89; \quad K_{FE} = K_{ED} = 5.06; \quad K_{AF} = K_{BE} = K_{CD} = 4.05$$

Step 2: Assume initial flows in each pipe as shown below

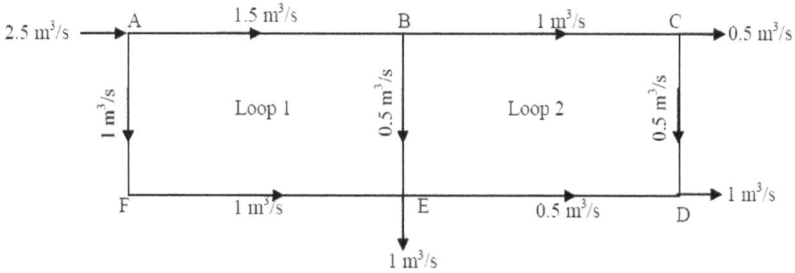

Step 3: Consider loop 1 and calculate ΔQ according to eqn. earlier; n = 2. Consider anticlockwise flows positive.

$$\Delta Q_1 = -\frac{1}{2}\frac{K_{AF}Q_{AF}^2 + K_{FE}Q_{FE}^2 - K_{BE}Q_{BE}^2 - K_{AB}Q_{AB}^2}{K_{AF}Q_{AF} + K_{FE}Q_{FE} - K_{BE}Q_{BE} - K_{AB}Q_{AB}}$$

$$= -\frac{1}{2}\frac{4.05\times1^2 + 5.06\times1^2 - 4.05\times0.5^2 - 3.88\times1.5^2}{4.05\times1 + 5.06\times1 + 4.05\times0.5 + 3.88\times1.5}$$

$$= 0.0187$$

Step 4: Consider loop 2 and calculate ΔQ

$$\Delta Q_2 = \frac{1}{2}\frac{4.05\times0.5^2 + 5.06\times0.5^2 - 4.05\times0.5^2 - 6.89\times1^2}{4.05\times0.5 + 5.06\times0.5 + 4.05\times0.5 + 6.89\times1.5}$$

$$= 0.2088$$

Step 5: Flows for next iteration

$Q_{AF} = 1 + 0.0187 = 1.0187$

$Q_{FE} = 1 + 0.0187 = 1.0187$

$Q_{BE} = 0.5 - 0.0187 + 0.2088 = 0.6901$

$Q_{AB} = 1.5 - 0.0187 = 1.4813$

$Q_{ED} = 0.5 + 0.2088 = 0.7088$

$Q_{CD} = 0.5 - 0.2088 = 0.2912$

$Q_{BC} = 1 - 0.2088 = 0.7912$

Repeat steps 2 to 5 with new flows till ΔQ is insignificant.

Linear Theory Method

Linear theory is more efficient when compared to Hardy-Cross method. Here we will demonstrate through the example network given below.

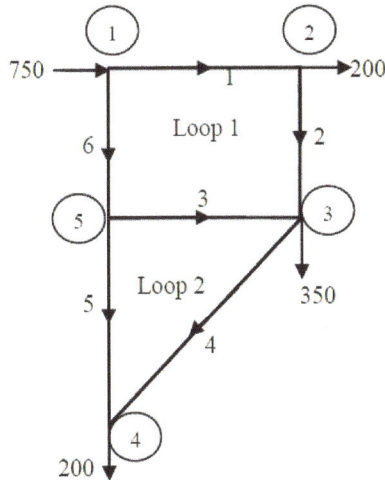

The equations for linear theory are:

Conservation of mass:

Node 1: $Q_1 + Q_6 = 750$

Node 2: $Q_1 - Q_2 = 200$

Node 3: $Q_2 + Q_3 - Q_4 = 350$

Node 4: $Q_4 + Q_5 = 200$

Node 5: $Q_6 - Q_3 - Q_5 = 0$

Among these 5 eqns. only 4 need to be used to avoid redundancy

Conservation of energy:

$$Loop\,1 = K_{12}Q_{12}^2 + K_{23}Q_{23}^2 - K_{35}Q_{35}^2 - K_{51}Q_{51}^2 = 0$$
$$Loop\,2 = K_{53}Q_{53}^2 + K_{34}Q_{34}^2 - K_{45}Q_{45}^2 = 0$$

Linearising the above eqns. using k=K_pQ, the eqns. can be written as

$$Loop\,1 = k_{12}Q_{12} + k_{23}Q_{23} - k_{35}Q_{35} - k_{51}Q_{51} = 0$$
$$Loop\,2 = k_{53}Q_{53} + k_{34}Q_{34} - k_{45}Q_{45} = 0$$

These 2 eqns. along with the 4 mass conservation eqns. can be solved to obtain 5 unknown discharges.

Optimization of Water Distribution Systems

Simulation of distribution networks as discussed above helps to determine the hydraulic parameters such as pressure heads, tank levels etc. These models are unable to determine the optimal or minimum cost system. In addition to the cost minimization, the typical goals of water distribution systems problem in designing pipe system can be:

A) Meeting the household demands.

B) Meeting the required water pressure at all nodes of the distribution system.

C) Optimal positioning of valves.

Therefore, designing water distribution system is a multiobjective problem, which is also characterized by nonlinearity resulting from the simulation model.

Since the main purpose of a water distribution system is to supply according to the demands with adequate pressure, a typical optimization problem will be to minimize the system's cost while meeting the demands at required pressures. Hence optimization problem can be stated as:

Minimize : Total cost (Capital cost + Energy cost for pumping water throughout the system)

Subject to:

(i) Hydraulic constraints

(ii) Water demand constraints

(iii) Pressure requirements

Groundwater

Groundwater management deals with planning, implementation, development and operation of water resources containing groundwater. Numerical-simulation models have been used extensively for understanding the flow characteristics of aquifers and evaluate the groundwater resource. While simulation models attempt various scenarios to find the best objective, optimization models directly consider the management objective taking care of all the constraints. In this lesson, we will discuss about the governing equations in these modes and various management models.

Groundwater Hydrology

Subsurface water is stored underneath in subsurface formations called aquifers. In an unconfined aquifer, the upper surface is the water table itself. On the other hand, a confined aquifer is confined under pressure greater than the atmospheric. A confined aquifer may be confined between two impermeable layers. An aquifer serves two functions: storage and transmission.

Storage function is exhibited through porosity Φ, specific yield S_y and storage coefficient S. Transmission function is exhibited through the permeability property (coefficient of permeability K). Porosity is the measure of the amount of water an aquifer can hold. Specific yield is the water drained from a saturated sample of unit volume. Specific retention is the water retained in the unit volume. Porosity is the sum of specific yield and specific retention. Storage coefficient is the volume of water an aquifer releases or stores per unit surface area per unit decline of head.

Permeability is a measure of the ease of movement of water through aquifers. The coefficient of permeability or hydraulic conductivity is the rate of flow of water through a unit cross-sectional area under a unit hydraulic gradient. Transmissivity, T is the rate of flow of water through a vertical strip of unit width extending the saturated thickness of the aquifer under a unit hydraulic gradient. Therefore,

$T = K\,b$ for a confined aquifer where b is the saturated thickness of aquifer

$T = K\,h$ for a confined aquifer where h is the head (saturated thickness)

Darcy's Law

The flow thorough an aquifer is expressed by Darcy's law which states that flow rate through a porous media is proportional to the head loss and inversely proportional to the length of flow path. It can be expressed as

$$v = -K\frac{\partial h}{\partial l} \qquad (1)$$

where v is the velocity or specific discharge, l is the length of flow along the average direction and $\frac{\partial h}{\partial l}$ is the rate of headloss per unit length. Then, the total discharge, q is

$$q = Av = -KA\frac{\partial h}{\partial l} \qquad (2)$$

Simulation of Groundwater Systems

Governing equations:

Darcy's law in terms of transmissivity is

$$v = -\frac{T}{b}\frac{\partial h}{\partial l} \quad \text{for confined aquifers} \tag{3}$$

$$v = -\frac{T}{h}\frac{\partial h}{\partial l} \quad \text{for unconfined aquifers} \tag{4}$$

Considering a two-dimensional horizontal flow as shown by a rectangular control volume element in the given figure, the general equations of flow can be expressed as:

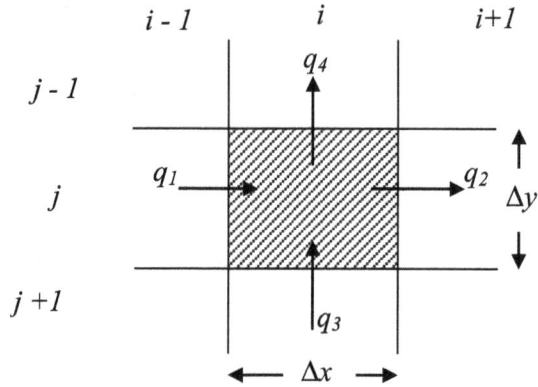

The flow discharge q = Av for four sides

$$q_1 = -T_x \Delta y \left(\frac{\partial h}{\partial x}\right)_1 \tag{5}$$

$$q_2 = -T_x \Delta y \left(\frac{\partial h}{\partial x}\right)_2 \tag{6}$$

$$q_3 = -T_y \Delta x \left(\frac{\partial h}{\partial x}\right)_3 \tag{7}$$

$$q_4 = -T_y \Delta x \left(\frac{\partial h}{\partial x}\right)_4 \tag{8}$$

where $A = \Delta x . h$ for unconfined case or $A = \Delta x.b$ for confined case and assuming constant transmissivities along the x and y directions. $\left(\frac{\partial h}{\partial x}\right)_1, \left(\frac{\partial h}{\partial x}\right)_2$ ... are the hydraulic gradients at sides 1,2 ,... respectively.

The rate at which water is stored or released in the element is

$$q_5 = S\Delta x \Delta y \left(\frac{\partial h}{\partial t}\right)_4 \tag{9}$$

where S is the storage coefficient of the element.

The flow rate for constant net withdrawal or recharge for time Δt.

$$q_6 = q_t \qquad (10)$$

Applying continuity law,

$$q_1 - q_2 + q_3 + q_4 = q_5 + q_6 \qquad (11)$$

Substituting eqns. 5 – 10 in above eqn, and dividing by Δx Δy and simplifying, the final form of eqn. 11 will be

$$T_x \frac{\partial^2 h}{\partial x^2} + T_y \frac{\partial^2 h}{\partial y^2} = S \frac{\partial h}{\partial t} + W \qquad (12)$$

Where $W = q / \Delta x \Delta y$

These equations can be written in finite difference form and solved for each rectangular element. The partial derivatives in eqns. 5-9 can be expressed in finite difference form as,

$$\left(\frac{\partial h}{\partial x}\right)_1 = \left(\frac{h_{i-1,j,t} - h_{i,j,t}}{\Delta x_i}\right)$$

$$\left(\frac{\partial h}{\partial x}\right)_2 = \left(\frac{h_{i,j,t} - h_{i+1,j,t}}{\Delta x_i}\right)$$

$$\left(\frac{\partial h}{\partial y}\right)_3 = \left(\frac{h_{i,j,t} - h_{i,j,t}}{\Delta y_j}\right) \qquad (13)$$

$$\left(\frac{\partial h}{\partial y}\right)_4 = \left(\frac{h_{i,j,t} - h_{i,j-1,t}}{\Delta y_j}\right)$$

$$\left(\frac{\partial h}{\partial t}\right) = \left(\frac{h_{i,j,t} - h_{i,j,t-1}}{\Delta t}\right)$$

These can be substituted in eqn. 12 and solved using finite element methods.

Optimization Model

Optimization models for hydraulic management for groundwater have been developed based on three approaches: embedding approach, optimal control approach and unit response matrix approach. In embedding approach, the equations from a simulation model are incorporated into an optimization model directly. In optimal control approach, the simulation model solves the governing equations, for each iteration of the optimization. It works on the concept of optimal control theory. In response matrix approach, a unit response matrix is generated by running the simulation model several times with unit pumpage at a single node. Total drawdowns are then determined by superpositions. We will discuss only about the embedding approach for steady-state one-dimensional confined and unconfined aquifers.

Steady State One-dimensional Confined Aquifer

Consider a confined aquifer with penetrating wells and flow in one-dimension as shown in the figure given below.

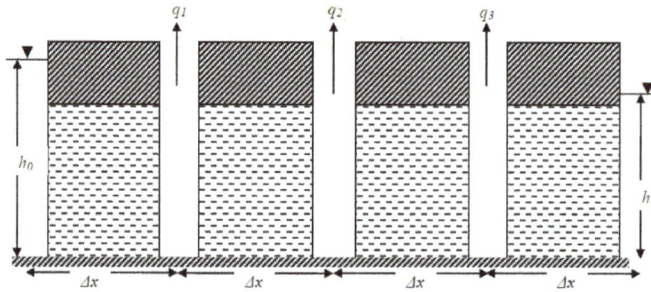

From eqn. 12, the governing equation is

$$\frac{\partial^2 h}{\partial x^2} = \frac{W}{T_x} \qquad \text{where} \quad \frac{\partial h}{\partial t} = 0 \qquad (14)$$

This can be expressed in finite difference form as (using central difference scheme)

$$\frac{h_{i+1} - 2h_i + h_{i-1}}{\Delta x^2} = \frac{W_i}{T_x} \qquad (15)$$

The optimization problem can be stated as

$$\text{Maximize } Z = \sum_i h_i \qquad (16)$$

where i is the number of wells.

Subject to

$$\sum_i W_i \geq W_{\min}$$

$$h_i \geq 0 \qquad (17)$$

$$W_i \geq 0$$

where W_{min} is the minimum production rate for each well. The pumpage can be finally determined from the relation $q_i = W.\Delta x_i^2$.

Example:

Formulate an LP model for the above confined aquifer for maximum heads. The boundaries have a constant head h_0 and h_5. The distance between the wells is Δx.

Solution:

Objective function: Maximize $Z = h_1 + h_2 + h_3 + h_4$

Subject to:

Acc. to eqn. 15

$$-2h_1 + h_2 - \frac{\Delta x^2}{T} W_1 = -h_0$$

$$h_1 - 2h_2 + h_3 - \frac{\Delta x^2}{T} W_2 = 0$$

$$h_2 - 2h_3 + h_4 - \frac{\Delta x^2}{T} W_3 = 0$$

$$h_3 - 2h_4 - \frac{\Delta x^2}{T} W_4 = -h_5$$

And acc. to eqn. 17

$$W_1 + W_2 + W_3 + W_4 \geq W_{min}$$
$$h_i \geq 0 \qquad i = 1,...,4$$
$$W_i \geq 0 \qquad i = 1,...,4$$

The unknowns are h_1, h_2, h_3, h_4 and W_1, W_2, W_3, W_4.

Steady State One-dimensional Unconfined Aquifer

The governing equation can be written as

$$\frac{\partial}{\partial x}\left(T \frac{\partial}{\partial x} \right) \quad W \qquad\qquad (18)$$

$$\frac{\partial^2 h^2}{\partial x} \quad \frac{W}{K}$$

where $T = Kh$. Substituting $w = h^2$ and assuming K is constant, the finite difference form can be written as

$$\frac{\partial}{\partial x^2} = \frac{w_{i+1} - 2w_i + w_{i-1}}{\Delta x^2} = \frac{2W_i}{K} \qquad\qquad (19)$$

The optimization problem is to

$$\text{Maximize } Z = \sum_i h_i \qquad\qquad (20)$$

where i is the number of wells.

Subject to

$$\sum_i W_i \geq W_{min}$$
$$w_i \geq 0 \qquad\qquad (21)$$
$$W_i \geq 0$$

After solving the above problem, the heads $h_i = \sqrt{w_i}$.

Example

Formulate a LP model to determine the steady state pumpages of an unconfined aquifer shown below.

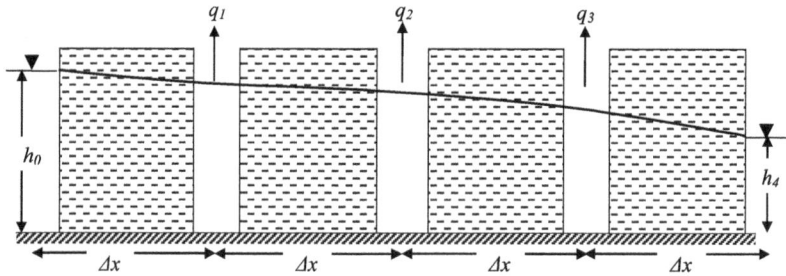

Solution

The optimization problem is

Objective function: Maximize $Z = w_1 + w_2 + w_3 + w_4$

Subject to:

Acc. to eqn. 19

$$- w_1 + w_2 - \text{———} W_1 = -w_0$$

$$w_1 - 2w_2 + w_3 - \text{———} W_2 = 0$$

$$w_1 - 2w_3 + w_4 - \text{———} W_3 = 0$$

$$w_3 - w_4 - \text{———} W_4 = -w_5$$

And acc. to eqn. 21

$$W_1 + W_2 + W_3 + W_4 \geq W_{min}$$
$$w_i \geq 0 \qquad i = 1, ..., 4$$
$$W_i \geq 0 \qquad i = 1, ..., 4$$

The unknowns are w_1, w_2, w_3, w_4 and W_1, W_2, W_3, W_4.

Flood Control

A weir was built on the Humber River (Ontario) to prevent a recurrence of a catastrophic flood.

Flood control refers to all methods used to reduce or prevent the detrimental effects of flood waters. Flood relief refers to methods used to reduce the effects of flood waters or high water levels.

Causes of Floods

Floods are caused by many factors (or a combination of any of these): heavy rainfall, highly accelerated snowmelt, severe winds over water, unusual high tides, tsunamis, or failure of dams, levees, retention ponds, or other structures that retained the water. Flooding can be exacerbated by increased amounts of impervious surface or by other natural hazards such as wildfires, which reduce the supply of vegetation that can absorb rainfall.

Periodic floods occur on many rivers, forming a surrounding region known as the flood plain.

During times of rain, some of the water is retained in ponds or soil, some is absorbed by grass and vegetation, some evaporates, and the rest travels over the land as surface runoff. Floods occur when ponds, lakes, riverbeds, soil, and vegetation cannot absorb all the water. Water then runs off the land in quantities that cannot be carried within stream channels or retained in natural ponds, lakes, and man-made reservoirs. About 30 percent of all precipitation becomes runoff and that amount might be increased by water from melting snow. River flooding is often caused by heavy rain, sometimes increased by melting snow. A flood that rises rapidly, with little or no warning, is called a flash flood. Flash floods usually result from intense rainfall over a relatively small area, or if the area was already saturated from previous precipitation.

Severe Winds Over Water

Even when rainfall is relatively light, the shorelines of lakes and bays can be flooded by severe winds—such as during hurricanes—that blow water into the shore areas.

Unusual High Tides

Coastal areas are sometimes flooded by unusually high tides, such as spring tides, especially when compounded by high winds and storm surges.

Effects of Floods

Flooding has many impacts. It damages property and endangers the lives of humans and other species. Rapid water runoff causes soil erosion and concomitant sediment deposition elsewhere (such as further downstream or down a coast). The spawning grounds for fish and other wildlife habitats can become polluted or completely destroyed. Some prolonged high floods can delay traffic in areas which lack elevated roadways. Floods can interfere with drainage and economical use of lands, such as interfering with farming. Structural damage can occur in bridge abutments, bank lines, sewer lines, and other structures within floodways. Waterway navigation and hydroelectric power are often impaired. Financial losses due to floods are typically millions of dollars each year, with the worst floods in recent U.S. history having cost billions of dollars.

Benefits of Flooding

There are many disruptive effects of flooding on human settlements and economic activities. However, flooding can bring benefits, such as making soil more fertile and providing nutrients in which it is deficient. Periodic flooding was essential to the well-being of ancient communities along the Tigris-Euphrates Rivers, the Nile River, the Indus River, the Ganges and the Yellow River, among others. The viability for hydrologically based renewable sources of energy is higher in flood-prone regions.

Protection and Control of Floods

Some methods of flood control have been practiced since ancient times. These methods include planting vegetation to retain extra water, terracing hillsides to slow flow downhill, and the construction of floodways (man-made channels to divert floodwater).

Other techniques include the construction of levees, lakes, dams, reservoirs, retention ponds to hold extra water during times of flooding.

Methods of Detection

This is the method used for remote sensing the disasters. Detection of disasters such as floods, earthquakes, and explosions are quite complex in previous days and range of detection is inappropriate. But, it came to possibilities by using Multi temporal visualization of Synthetic Aperture Radar (SAR) images. But to obtain the good SAR images perfect spatial registration and very precise calibration are necessary to specify changes that have occurred. Calibration of SAR is very complex and also a sensitive problem. Possibly errors may occur after calibration that involves data fusion and visualization process. Traditional image pre-processing cannot be used here due to the on-Gaussian of radar back scattering, but a processing method called "cross calibration/normalization" is used to solve this problem. The application generates a single disaster image called "fast-ready disaster map" from multitemporal SAR images. These maps are generated without user interaction and helps in providing immediate first aid to the people. This process also provides image enhancement and comparison between numerous images using data fusion and visualization process. This proposed processing includes filtering, histogram truncation and equalization steps. The process also helps in identifying the permanent waters and other classes by combined composition of pre-disaster and post-disaster images into a color image for better identity.

Methods of Controlling Floods

Dams

Many dams and their associated reservoirs are designed completely or partially to aid in flood protection and control. Many large dams have flood-control reservations in which the level of a reservoir must be kept below a certain elevation before the onset of the rainy/summer melt season to allow a certain amount of space in which floodwaters can fill. The term dry dam refers to a dam that serves purely for flood control without any conservation storage (e.g. Mount Morris Dam, Seven Oaks Dam).

Water-Gate

The Water-Gate Flood barrier is a rapid response barrier which can be rolled out in a matter of minutes. It is unique in the way that it self deploys using the weight of water to hold it back. The product has been FM Approved following testing from the US Army. It is used in 30 countries around the world, and notably by the Environment Agency in the UK.

Diversion Canals

Floods can be controlled by redirecting excess water to purpose-built canals or floodways, which in turn divert the water to temporary holding ponds or other bodies of water where there is a lower risk or impact to flooding. Examples of flood control channels include the Red River Floodway that protects the City of Winnipeg (Canada) and the Manggahan Floodway that protects the City of Manila (Philippines).

Self-closing Flood Barrier

The self-closing flood barrier (SCFB) is a flood defense system designed to protect people and property from inland waterway floods caused by heavy rainfall, gales or rapid melting snow. The SCFB can be built to protect residential properties and whole communities, as well as industrial or other strategic areas. The barrier system is constantly ready to deploy in a flood situation, it can be installed in any length and uses the rising flood water to deploy. Barrier systems have already been built and installed in Belgium, Italy, Ireland, the Netherlands, Thailand, United Kingdom, Vietnam, Australia, Russia and the United States. Millions of documents at the National Archives building in Washington DC are protected by two SCFBs.

River Defences

In many countries, rivers are prone to floods and are often carefully managed. Defenses such as levees, bunds, reservoirs, and weirs are used to prevent rivers from bursting their banks. When these defenses fail, emergency measures such as sandbags, hydrosacks or portable inflatable tubes are used.

A weir, also known as a lowhead dam, is most often used to create millponds, but on the Humber River in Toronto, a weir was built near Raymore Drive to prevent a recurrence of the flood damage caused by Hurricane Hazel in October 1954.

Coastal Defenses

Coastal flooding has been addressed in Europe and the Americas with coastal defences, such as sea walls, beach nourishment, and barrier islands.

Tide gates are used in conjunction with dykes and culverts. They can be placed at the mouth of streams or small rivers, where an estuary begins or where tributary streams, or drainage ditches connect to sloughs. Tide gates close during incoming tides to prevent tidal waters from moving upland, and open during outgoing tides to allow waters to drain out via the culvert and into the estuary side of the dike. The opening and closing of the gates is driven by a difference in water level on either side of the gate.

Temporary Perimeter Barriers

In 1988, a method of using water to control was discovered. This was accomplished by containing 2 parallel tubes within a third outer tube. When filled, this structure formed a non-rolling wall of water that can control 80 percent of its height in external water depth, with dry ground behind it. Eight foot tall water filled barriers were used to surround Fort Calhoun Nuclear Generating Station during the 2011 Missouri River Flooding. Instead of trucking in sandbag material for a flood, stacking it, then trucking it out to a hazmat disposal site, flood control can be accomplished by using the on site water. However, these are not fool proof. A 8 feet (2.4 m) high 2,000 feet (610 m) long water filled rubber flood berm that surrounded portions of the plant was punctured by a skid-steer loader and it collapsed flooding a portion of the facility.

In 1999, A group of Norwegian Engineers founded and patented Aquafence. A transportable, removable, and reusable flood barrier which uses the water's weight against itself. In 2013, AquaFence was awarded the highest level USA ANSI Certification after more than one year of testing of the system by US ARMY Corps of Engineers as well as parts testing and production review by FM Global. Both commercial and municipal customers spanning across The United States of America, Europe and Asia. In the US alone, AquaFence removable flood panels are protecting more than $10 billion worth of real estate as well as cities and public utilities.

Property Level Protection

Property level protection is advocated as a method of trying to manage flood risk at the receptor (at a property or community scale).

Flood Control by Continent

Americas

An elaborate system of flood way defenses can be found in the Canadian province of Manitoba. The Red River flows northward from the United States, passing through the city of Winnipeg (where it meets the Assiniboine River) and into Lake Winnipeg. As is the case with all north-flowing rivers in the temperate zone of the Northern Hemisphere, snow melt in southern sections may cause river levels to rise before northern sections have had a chance to completely thaw. This can lead to devastating flooding, as occurred in Winnipeg during the spring of 1950. To protect the city from future floods, the Manitoba government undertook the construction of a massive system of diversions, dikes, and flood ways (including the Red River Flood way and the Portage Diversion). The system kept Winnipeg safe during the 1997 flood which devastated many communities upriver from Winnipeg, including Grand Forks, North Dakota and Ste. Agathe, Manitoba.

In the United States, the U.S. Army Corps of Engineers is the lead flood control agency. After Hurricane Sandy, New York City's Metropolitan Transportation Authority (MTA) initiated multiple flood barrier projects to protect the transit assets in Manhattan. In one case, the MTA's New York City Transit Authority (NYCT) sealed subway entrances in lower Manhattan using a deployable fabric cover system called Flex-Gate, a system that protects the subway entrances against 14 feet (4.3 m) of water. Extreme storm flood protection levels have been revised based on new Federal Emergency Management Agency guidelines for 100-year and 500-year design flood elevations.

In the New Orleans Metropolitan Area, 35 percent of which sits below sea level, is protected by hundreds of miles of levees and flood gates. This system failed catastrophically, with numerous breaks, during Hurricane Katrina (2005) in the city proper and in eastern sections of the Metro Area, resulting in the inundation of approximately 50 percent of the metropolitan area, ranging from a few inches to twenty feet in coastal communities.

The Morganza Spillway provides a method of diverting water from the Mississippi River when a river flood threatens New Orleans, Baton Rouge and other major cities on the lower Mississippi. It is the largest of a system of spillways and floodways along the Mississippi. Completed in 1954, the spillway has been opened twice, in 1973 and in 2011.

In an act of successful flood prevention, the federal government offered to buy out flood-prone properties in the United States in order to prevent repeated disasters after the 1993 flood across the Midwest. Several communities accepted and the government, in partnership with the state, bought 25,000 properties which they converted into wetlands. These wetlands act as a sponge in storms and in 1995, when the floods returned, the government did not have to expend resources in those areas.

Asia

In India, Bangladesh and China, flood diversion areas are rural areas that are deliberately flooded in emergencies in order to protect cities.

The consequences of deforestation and changing land use on the risk and severity of flooding are subjects of discussion. In assessing the impacts of Himalayan deforestation on the Ganges-Brahmaputra Lowlands, it was found that forests would not have prevented or significantly reduced flooding in the case of an extreme weather event. However, more general or overview studies agree on the negative impacts that deforestation has on flood safety - and the positive effects of wise land use and reforestation.

Many have proposed that loss of vegetation (deforestation) will lead to an increased risk of flooding. With natural forest cover the flood duration should decrease. Reducing the rate of deforestation should improve the incidents and severity of floods.

Africa

In Egypt, both the Aswan Dam (1902) and the Aswan High Dam (1976) have controlled various amounts of flooding along the Nile river.

Europe

Following the misery and destruction caused by the 1910 Great Flood of Paris, the French government built a series of reservoirs called Les Grands Lacs de Seine (or Great Lakes) which helps remove pressure from the Seine during floods, especially the regular winter flooding.

Flood blocking the road in Jerusalem

London is protected from flooding by a huge mechanical barrier across the River Thames, which is raised when the water level reaches a certain point.

Venice has a similar arrangement, although it is already unable to cope with very high tides. The defenses of both London and Venice will be rendered inadequate if sea levels continue to rise.

The largest and most elaborate flood defenses can be found in the Netherlands, where they are referred to as Delta Works with the Oosterschelde dam as its crowning achievement. These works were built in response to the North Sea flood of 1953, in the southwestern part of the Netherlands. The Dutch had already built one of the world's largest dams in the north of the country. The Afsluitdijk closing occurred in 1932.

The Saint Petersburg Flood Prevention Facility Complex was completed in 2008, in Russia, to protect Saint Petersburg from storm surges. It also has a main traffic function, as it completes a ring road around Saint Petersburg. Eleven dams extend for 25.4 kilometres (15.8 mi) and stand 8 metres (26 ft) above water level.

Flood Clean-up Safety

Clean-up activities following floods often pose hazards to workers and volunteers involved in the effort. Potential dangers include electrical hazards, carbon monoxide exposure, musculoskeletal hazards, heat or cold stress, motor vehicle-related dangers, fire, drowning, and exposure to hazardous materials. Because flooded disaster sites are unstable, clean-up workers might encounter sharp jagged debris, biological hazards in the flood water, exposed electrical lines, blood or other body fluids, and animal and human remains. In planning for and reacting to flood disasters, managers provide workers with hard hats, goggles, heavy work gloves, life jackets, and watertight boots with steel toes and insoles.

Future

Europe is at the forefront of the flood control technology, with low-lying countries such as the Netherlands and Belgium developing techniques that can serve as examples to other countries facing similar problems.

After Hurricane Katrina, the US state of Louisiana sent politicians to the Netherlands to take a tour of the complex and highly developed flood control system in place in the Netherlands. With a BBC article quoting experts as saying 70 percent more people will live in delta cities by 2050, the number of people impacted by a rise in sea level will greatly increase. The Netherlands has one of the best flood control systems in the world and new ways to deal with water are constantly being developed and tested, such as the underground storage of water, storing water in reservoirs in large parking garages or on playgrounds, Rotterdam started a project to construct a floating housing development of 120 acres (0.49 km²) to deal with rising sea levels. Several approaches, from high-tech sensors detecting imminent levee failure to movable semi-circular structures closing an entire river, are being developed or used around the world. Regular maintenance of hydraulic structures, however, is another crucial part of flood control.

Urban Runoff

Urban runoff is surface runoff of rainwater created by urbanization. This runoff is a major source of flooding and water pollution in urban communities worldwide.

Impervious surfaces (roads, parking lots and sidewalks) are constructed during land development. During rain storms and other precipitation events, these surfaces (built from materials such as asphalt and concrete), along with rooftops, carry polluted stormwater to storm drains, instead of allowing the water to percolate through soil. This causes lowering of the water table (because groundwater recharge is lessened) and flooding since the amount of water that remains on the surface is greater. Most municipal storm sewer systems discharge stormwater, untreated, to streams, rivers and bays. This excess water can also make its way into people's properties through basement backups and seepage through building wall and floors.

Urban Flooding

Urban runoff is a major cause of urban flooding, the inundation of land or property in a built-up environment caused by rainfall overwhelming the capacity of drainage systems, such as storm sewers. Triggered by events such as flash flooding, storm surges, overbank flooding, or snow melt, urban flooding is characterized by its repetitive, costly and systemic impacts on communities, regardless of whether or not these communities are located within formally designated floodplains or near any body of water.

Urban runoff flowing into a storm drain

There are several ways in which stormwater enters properties: backup through sewer pipes, toilets and sinks into buildings; seepage through building walls and floors; the accumulation of water on property and in public rights-of-way; and the overflow of water from water bodies such as rivers and lakes. Where properties are built with basements, urban flooding is the primary cause of basement flooding.

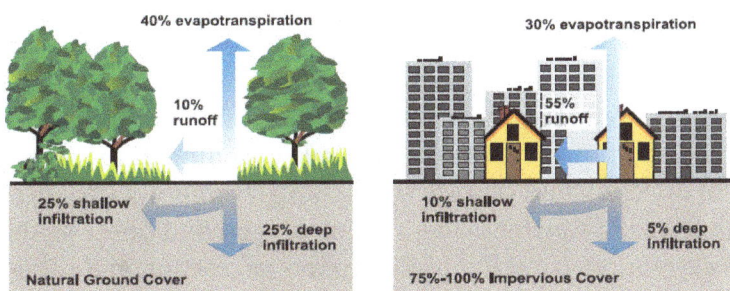

Relationship between impervious surfaces and surface runoff

Flood flows in urban environments have been investigated relatively recently despite many centuries of flood events. Some researchers mentioned the storage effect in urban areas. Several studies looked into the flow patterns and redistribution in streets during storm events and the implication in terms of flood modelling. Some recent research considered the criteria for safe evacuation of individuals in flooded areas. But some recent field measurements during the 2010–2011 Queensland floods showed that any criterion solely based upon the flow velocity, water depth or specific momentum cannot account for the hazards caused by the velocity and water depth fluctuations. These considerations ignore further the risks associated with large debris entrained by the flow motion.

Pollutants

Water running off these impervious surfaces tends to pick up gasoline, motor oil, heavy metals, trash and other pollutants from roadways and parking lots, as well as fertilizers and pesticides from lawns. Roads and parking lots are major sources of polycyclic aromatic hydrocarbons (PAHs), which are created as combustion byproducts of gasoline and other fossil fuels, as well as of the heavy metals nickel, copper, zinc, cadmium, and lead. Roof runoff contributes high levels of synthetic organic compounds and zinc (from galvanized gutters). Fertilizer use on residential lawns, parks and golf courses is a measurable source of nitrates and phosphorus in urban runoff when fertilizer is improperly applied or when turf is over-fertilized.

A creek filled with urban runoff after a storm

Eroding soils or poorly maintained construction sites can often lead to increased sedimentation in runoff. Sedimentation often settles to the bottom of water bodies and can directly affect water quality. Excessive levels of sediment in water bodies can increase the risk of infection and disease through high levels of nutrients present in the soil. These high levels of nutrients can reduce oxygen and boost algae growth while limiting native vegetation growth. Limited native vegetation and excessive algae has the potential to disrupt the entire aquatic ecosystem due to limited light penetration, lower oxygen levels, and reduced food reserves. Excessive levels of sediment and suspended solids have the potential to damage existing infrastructure as well. Sedimentation can increase runoff by plugging underground injection systems, thereby increasing the amount of runoff on the surface. Increased sedimentation levels can also reduce storage behind reservoirs. This reduction of reservoir capacities can lead to increased expenses for public land agencies while also impacting the quality of water recreational areas.

Runoff can also induce heavy metal poisoning in ocean life. Small amounts of heavy metals are carried by runoff into the oceans. These metals are ingested by ocean life. These heavy metals cannot be disposed so they accumulate within the animals. Over time, these metals build up to a toxic level, and the animal dies. This heavy metal poisoning can also affect humans. If we eat a poisoned animal, we have a chance of getting heavy metal poisoning too.

As stormwater is channeled into storm drains and surface waters, the natural sediment load discharged to receiving waters decreases, but the water flow and velocity increases. In fact, the impervious cover in a typical city creates five times the runoff of a typical woodland of the same size.

Effects

A 2008 report by the United States National Research Council (textbox below) identified urban runoff as a leading source of water quality problems.

...further declines in water quality remain likely if the land-use changes that typify more diffuse sources of pollution are not addressed... These include land-disturbing agricultural, silvicultural, urban, industrial, and construction activities from which hard-to-monitor pollutants emerge during wet-weather events. Pollution from these landscapes has been almost universally acknowledged as the most pressing challenge to the restoration of waterbodies and aquatic ecosystems nationwide.

– National Research Council, *Urban Stormwater Management in the United States*

Weasel Brook in Passaic, New Jersey has been channelized with concrete walls to control localized flooding.

The runoff also increases temperatures in streams, harming fish and other organisms. (A sudden burst of runoff from a rainstorm can cause a fish-killing shock of hot water.) Also, road salt used to melt snow on sidewalks and roadways can contaminate streams and groundwater aquifers.

One of the most pronounced effects of urban runoff is on watercourses that historically contained little or no water during dry weather periods (often called *ephemeral streams*). When an area around such a stream is urbanized, the resultant runoff creates an unnatural year-round streamflow that hurts the vegetation, wildlife and stream bed of the waterway. Containing little or no sediment relative to the historic ratio of sediment to water, urban runoff rushes down the stream channel, ruining natural features such as meanders and sandbars, and creates severe erosion—increasing sediment loads at the mouth while severely incising the stream bed upstream. As an example, on many Southern California beaches at the mouth of a waterway, urban runoff carries trash, pollutants, excessive silt, and other wastes, and can pose moderate to severe health hazards.

Because of fertilizer and organic waste that urban runoff often carries, eutrophication often occurs in waterways affected by this type of runoff. After heavy rains, organic matter in the waterway is relatively high compared with natural levels, spurring growth of algae blooms that soon consume most of the oxygen. Once the naturally occurring oxygen in the water is depleted, the algae blooms die, and their decomposition causes further eutrophication. Algae blooms mostly occur in areas with still water, such as stream pools and the pools behind dams, weirs, and some drop structures. Eutrophication usually comes with deadly consequences for fish and other aquatic organisms.

Excessive stream bank erosion may cause flooding and property damage. For many years governments have often responded to urban stream erosion problems by modifying the streams through construction of hardened embankments and similar control structures using concrete and masonry materials. Use of these hard materials destroys habitat for fish and other animals. Such a project may stabilize the immediate area where flood damage occurred, but often it simply shifts the problem to an upstream or downstream segment of the stream.

Urban flooding has significant economic implications. In the US, industry experts estimate that wet basements can lower property values by 10%-25% and are cited among the top reasons for not purchasing a home. According to the U.S Federal Emergency Management Agency (FEMA), almost 40% of small businesses never reopen their doors following a flooding disaster. In the UK, urban flooding is estimated to cost £270 million a year in England and Wales; 80,000 homes are at risk.

A study of Cook County, Illinois, identified 177,000 property damage insurance claims made across 96% of the county's ZIP codes over a five-year period from 2007-2011. This is the equivalent of one in six properties in the County making a claim. Average payouts per claim were $3,733 across all types of claims, with total claims amounting to $660 million over the five years examined.

Despite concerted efforts, many communities lack the funds to fully address these issues, and often seek funds elsewhere. Numerous watersheds within Los Angeles County, California do not meet state water quality standards, despite spending $100 million a year on clean water programs to combat issues such as urban runoff. To combat this problem, officials have introduced a measure that would assess a fee to homeowners and local businesses in attempt to raise $290 million for effective urban runoff management.

Prevention and mitigation

Effective control of urban runoff involves reducing the velocity and flow of stormwater, as well as reducing pollutant discharges. A variety of stormwater management practices and systems may be used to reduce the effects of urban runoff. Some of these techniques (called best management practices (BMPs) in the US), focus on water quantity control, while others focus on improving water quality, and some perform both functions.

A percolation trench infiltrates stormwater through permeable soils into the groundwater aquifer.

Pollution prevention practices include low impact development (LID) or green infrastructure techniques - known as Sustainable Drainage Systems (SuDS) in the UK, and Water-Sensitive Urban Design (WSUD) in Australia and the Middle East - such as the installation of green roofs and improved chemical handling (e.g. management of motor fuels & oil, fertilizers and pesticides). Runoff mitigation systems include infiltration basins, bioretention systems, constructed wetlands, retention basins and similar devices.

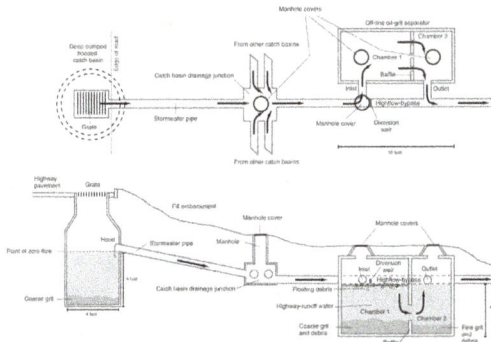

An oil-grit separator is designed to capture settleable solids, oil and grease, debris and floatables in runoff from roads and parking lots

Providing effective urban runoff solutions often requires proper city programs that take into account the needs and differences of the community. Factors such as a city's mean temperature, precipitation levels, geographical location, and airborne pollutant levels can all effect rates of pollution in urban runoff and present unique challenges for management. Human factors such as urbanization rates, land use trends, and chosen building materials for impervious surfaces often exacerbate these issues.

The implementation of citywide maintenance strategies such as street sweeping programs can also be an effective method in improving the quality of urban runoff. Street sweeping vacuums collect particles of dust and suspended solids often found in public parking lots and roads that often end up in runoff.

Blue drain and yellow fish symbol used by the UK Environment Agency to raise awareness of the ecological impacts of contaminating surface drainage

Educational programs can also be an effective tool for managing urban runoff. Local businesses and individuals can have an integral role in reducing pollution in urban runoff simply through their practices, but often are unaware of regulations. Creating a productive discussion on urban runoff and the importance of effective disposal of household items can help to encourage environmentally friendly practices at a reduced cost to the city and local economy.

Stormwater

Urban stormwater management systems are meant to guide, control and modify the quantity and quality of surface runoff. There are basically five subsystems which characterizes the urban drainage system: (i) surface runoff subsystem (2) storm sewer subsystem (3) detention subsystem (4) open channel transport subsystem and (5) receivers such as rivers, lakes or oceans. In this lesson we will discuss about the various subsystems and the design of storm sewers using various methods.

Subsystems

The surface runoff subsystem transforms the rainfall input into surface water runoff. The outputs runoff hydrograph from surface runoff subsystem is the input to the storm sewer subsystem. Storm sewer subsystem transports runoff to either a detention subsystem or an open channel transport subsystem or a receiver subsystem. Output releases from a detention subsystem can be the input to an open channel subsystem or a receiver subsystem. Output releases from open channel subsystem can be the input to a detention subsystem or a receiver subsystem.

In urban stromwater management, the determination of runoff yield and the optimal design of storm sewer networks are very important. Strom water runoff alleviation is a major task.

Storm Sewers

Storm sewers play an important role in urban stromwater management. A storm sewer system may consist of a number of sewers, junctions, manholes and inlets in addition to regulating and operating devices. The design of storm sewer includes determining the diameter, slopes and crown elevations of each pipe in the network. The design models can be divided as hydraulic design models and optimization design models. The hydraulic design models determine the sewer diameters by considering only the hydraulic parameters. In optimization design models, the minimum sewer size that is able to carry the design discharge under full pipe gravity conditions is determined. In these design models, the sewer system layout is predetermined and the sewer slope is assumed to be same as that of ground slope. The assumptions and constraints commonly used in storm sewer design are:

(i) Sewer is usually designed for gravity flow, No need of pumping stations or pressurized sewers.

(ii) The pipes used are commercially available circular ones with a minimum diameter of 20 cm.

(iii) The design diameter should the smallest available pipe with flow capacity equal or greater than the design discharge and satisfies all constraints.

(iv) The storm sewers must be placed well below the ground level to prevent frost, drain basements and also to allow sufficient cushioning against breakage due to ground surface loading. Therefore, minimum cover depths should be specified.

(v) At junctions, the crown elevation of the upstream sewer should not be lower than that of the downstream sewer.

(vi) A minimum permissible flow velocity at design discharge or at barely full pipe gravity flow should be specified to prevent excessive deposition of solids in the sewers.

(vii) A maximum permissible flow velocity should be specified to prevent scouring effects.

(viii) The downstream sewer should not be smaller than any of the upstream sewers at any junction.

(ix) The sewer system is a dendritic network converging towards downstream without any closed loops.

References

- "Parking Lot and Street Cleaning". National Menu of Stormwater Best Management Practices. EPA. Archived from the original on 2015-08-28. Retrieved 2014-12-24

- G. Allen Burton, Jr., Robert Pitt (2001). Stormwater Effects Handbook: A Toolbox for Watershed Managers, Scientists, and Engineers. New York: CRC/Lewis Publishers. ISBN 0-87371-924-7

- Werner, MGF; Hunter, NM; Bates, PD (2006). "Identifiability of Distributed Floodplain Roughness Values in Flood Extent Estimation". Journal of Hydrology. 314: 139–157. doi:10.1016/j.jhydrol.2005.03.012

- Water Environment Federation, Alexandria, VA; and American Society of Civil Engineers, Reston, VA. "Urban Runoff Quality Management." WEF Manual of Practice No. 23; ASCE Manual and Report on Engineering Practice No. 87. 1998. ISBN 1-57278-039-8

- National Research Council (United States) (2009). Urban Stormwater Management in the United States (Report). Washington, D.C.: National Academies Press. p. 24. doi:10.17226/12465. ISBN 978-0-309-12539-0

- Laws, Edward A.; Roth, Lauren (2004). "Impact of Stream Hardening on Water Quality and Metabolic Characteristics of Waimanalo and Kane'ohe Streams, O'ahu, Hawaiian Islands". Pacific Science. University of Hawai'i Press. 58 (2). doi:10.1353/psc.2004.0019. hdl:10125/2725. ISSN 0030-8870

3

Impact of Irrigation on Water Resources

Irrigation adversely impacts soil and water and contributes to their pollution. Its direct effects include increased evaporation, decline in river flow, and also affect moisture, and other environmental factors of the area. Indirect effects of irrigation are soil salination, waterlogging, saltwater intrusion etc. The chapter serves as a source to understand the major categories related to impacts of irrigation.

Soil and Water for Plant

Both soil and water are essential for plant growth. The soil provides a structural base to the plants and allows the root system (the foundation of the plant) to spread and get a strong hold. The pores of the soil within the root zone hold moisture which clings to the soil particles by surface tension in the driest state or may fill up the pores partially or fully saturating with it useful nutrients dissolved in water, essential for the growth of the plants. The roots of most plants also require oxygen for respiration. Hence, full saturation of the soil pores leads to restricted root growth for these plants. (There are exceptions, though, like the rice plant, in which the supply of oxygen to the roots is made from the leaves through aerenchyma cells which are continuous from the leaves to the roots).

Since irrigation practice is essentially, an adequate and timely supply of water to the plant root zone for optimum crop yield, the study of the inter relationship between soil pores, its water-holding capacity and plant water absorption rate is fundamentally important. Though a study in detail would mostly be of importance to an agricultural scientist, in this lesson we discuss the essentials which are important to a water resources engineer contemplating the development of a command area through scientifically designed irrigation system.

Soil-water System

Soil is a heterogeneous mass consisting of a three phase system of solid, liquid and gas. Mineral matter, consisting of sand, silt and clay and organic matter form the largest fraction of soil and serves as a framework (matrix) with numerous pores of various proportions. The void space within the solid particles is called the soil pore space. Decayed organic matter derived from the plant and animal remains are dispersed within the pore space. The soil air is totally expelled from soil when water is present in excess amount than can be stored.

On the other extreme, when the total soil is dry as in a hot region without any sup-ply of water either naturally by rain or artificially by irrigation, the water molecules surround the soil particles as a thin film. In such a case, pressure lower than atmo-spheric thus results due to surface tension capillarity and it is not possible to drain out the water by gravity. The salts present in soil water further add to these forces by way of osmotic pressure. The roots of the plants in such a soil state need to exert at least an equal amount of force for extracting water from the soil mass for their growth.

In the following sections, we discuss certain important terms and concepts related to the soil-water relations. First, we start with a discussion on soil properties and types of soils.

Soil Properties

Soil is a complex mass of mineral and organic particles. The important properties that classify soil according to its relevance to making crop production (which in turn affects the decision making process of irrigation engineering) are:

- Soil texture

- Soil structure

Soil Texture

This refers to the relative sizes of soil particles in a given soil. According to their sizes, soil particles are grouped into gravel, sand, silt and day. The relative proportions of sand, silt and clay is a soil mass determines the soil texture. Figure below presents the textural classification of 12 main classes as identified by the US department of agricul-ture, which is also followed by the soil survey organizations of India.

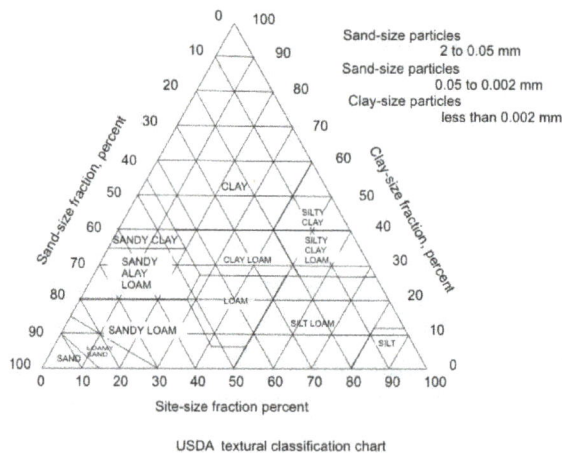

USDA textural classification chart

According to textural gradations a soil may be broadly classified as:

- Open or light textural soils: these are mainly coarse or sandy with low content of silt and clay.

- Medium textured soils: these contain sand, silt and clay in sizeable proportions, like loamy soil.

- Tight or heavy textured soils: these contain high proportion of clay.

Soil structure:

This refers to the arrangement of soil particles and aggregates with respect to each other. Aggregates are groups of individual soil particles adhering together. Soil structure is recognized as one of the most important properties of soil mass, since it influences aeration, permeability, water holding capacity, etc. The classification of soil structure is done according to three indicators as:-

- Type: there are four types of primary structures-platy, prism-like, block like and spheroidal.

- Class: there are five recognized classes in each of the primary types. These are very fine, fine, medium, coarse and very coarse.

- Grade: this represents the degree of aggradation that is the proportion between aggregate and unaggregated material that results when the aggregates are displaced or gently crushed. Grades are termed as structure less, weak, moderate, strong and very strong depending on the stability of the aggregates when disturbed.

Soil Classification

Soil types

Soil classification deals with the systematic categorization of soils based on distinguishing characteristics as well as criteria that dictate choices in use.

Overview

Soil classification is a dynamic subject, from the structure of the system itself, to the definitions of classes, and finally in the application in the field. Soil classification can be approached from the perspective of soil as a material and soil as a resource.

Engineering

Engineers, typically geotechnical engineers, classify soils according to their engineering properties as they relate to use for foundation support or building material. Modern engineering classification systems are designed to allow an easy transition from field observations to basic predictions of soil engineering properties and behaviors.

The most common engineering classification system for soils in North America is the Unified Soil Classification System (USCS). The USCS has three major classification groups: (1) coarse-grained soils (e.g. sands and gravels); (2) fine-grained soils (e.g. silts and clays); and (3) highly organic soils (referred to as "peat"). The USCS further subdivides the three major soil classes for clarification. It distinguishes sands from gravels by grain size, and further classifying some as "well-graded" and the rest as "poorly-graded". Silts and clays are distinguished by the soils' Atterberg limits, and separates "high-plasticity" from "low-plasticity" soils as well. Moderately organic soils are considered subdivisions of silts and clays, and are distinguished from inorganic soils by changes in their plasticity properties (and Atterberg limits) on drying. The European soil classification system (ISO 14688) is very similar, differing primarily in coding and in adding an "intermediate-plasticity" classification for silts and clays, and in minor details.

Other engineering soil classification systems in the United States include the AASHTO Soil Classification System, which classifies soils and aggregates relative to their suitability for pavement construction, and the Modified Burmister system, which works similarly to the USCS, but includes more coding for various soil properties.

A full geotechnical engineering soil description will also include other properties of the soil including color, in-situ moisture content, in-situ strength, and somewhat more detail about the material properties of the soil than is provided by the USCS code. The USCS and additional engineering description is standardized in ASTM D 2487.

Soil Science

For soil resources, experience has shown that a natural system approach to classification, i.e. grouping soils by their intrinsic property (soil morphology), behaviour, or genesis, results in classes that can be interpreted for many diverse uses. Differing concepts of pedogenesis, and differences in the significance of morphological features to various land uses can affect the classification approach. Despite these differences, in a well-constructed system, classification criteria group similar concepts so that interpretations do not vary widely. This is in contrast to a technical system approach to soil classification, where soils are grouped according to their fitness for a specific use and their edaphic characteristics.

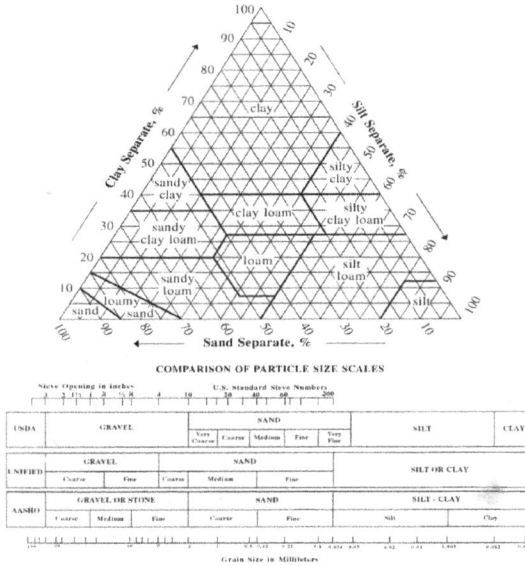

Soil texture triangle showing the USDA classification system based on grain size

Natural system approaches to soil classification, such as the French Soil Reference System (Référentiel pédologique français) are based on presumed soil genesis. Systems have developed, such as USDA soil taxonomy and the World Reference Base for Soil Resources, which use taxonomic criteria involving soil morphology and laboratory tests to inform and refine hierarchical classes.

Another approach is numerical classification, also called ordination, where soil individuals are grouped by multivariate statistical methods such as cluster analysis. This produces natural groupings without requiring any inference about soil genesis.

In soil survey, as practiced in the United States, soil classification usually means criteria based on soil morphology in addition to characteristics developed during soil formation. Criteria are designed to guide choices in land use and soil management. As indicated, this is a hierarchical system that is a hybrid of both *natural* and objective criteria. USDA soil taxonomy provides the core criteria for differentiating soil map units. This is a substantial revision of the 1938 USDA soil taxonomy which was a strictly natural system. Soil taxonomy based soil map units are additionally sorted into classes based on technical classification systems. Land Capability Classes, hydric soil, and prime farmland are some examples.

In addition to scientific soil classification systems, there are also vernacular soil classification systems. Folk taxonomies have been used for millennia, while scientifically based systems are relatively recent developments.

Soil Classifications for OSHA

The U.S. Occupational Safety and Health Administration (OSHA) requires the classification of soils to protect workers from injury when working in excavations and trenches. OSHA uses 3 soil classifications plus one for rock, based primarily on strength but also :

- Stable Rock: natural solid mineral matter that can be excavated with vertical sides and remain intact while exposed.

- Type A - cohesive, plastic soils with unconfined compressive strength greater than 1.5 ton per square foot (tsf)(144 kPa), and meeting several other requirements (with a lateral soil pressure of 25 psf per ft of depth)

- Type B - cohesive soils with unconfined compressive strength between 0.5 tsf (48 kPa) and 1.5 tsf (144 kPa), or unstable dry rock, or soils which would otherwise be Type A (with a lateral soil pressure of 45 psf per ft of depth)

- Type C - granular soils or cohesive soils with unconfined compressive strength less than 0.5 tsf (48 kPa) or any submerged or freely seeping soil or adversely bedded soils (with a lateral soil pressure of 80 psf per ft of depth)

- Type C60 - A subtype of Type C soil, though is not officially recognized by OSHA as a separate type, has a lateral soil pressure of 60 psf per ft of depth

Each of the soil classifications has implications for the way the excavation must be made or the protections (sloping, shoring, shielding, etc.) that must be provided to protect workers from collapse of the excavated bank.

Classification of Soil Water

As stated earlier, water may occur in the soil pores in varying proportions. Some of the definitions related to the water held in the soil pores are as follows:

- Gravitational water: A soil sample saturated with water and left to drain the excess out by gravity holds on to a certain amount of water. The volume of water that could easily drain off is termed as the gravitational water. This water is not available for plants use as it drains off rapidly from the root zone.

- Capillary water: the water content retained in the soil after the gravitational water has drained off from the soil is known as the capillary water. This water is held in the soil by surface tension. Plant roots gradually absorb the capillary water and thus constitute the principle source of water for plant growth.

- Hygroscopic water: the water that an oven dry sample of soil absorbs when exposed to moist air is termed as hygroscopic water. It is held as a very thin film over the surface of the soil particles and is under tremendous negative (gauge) pressure. This water is not available to plants.

The above definitions of the soil water are based on physical factors. Some properties of soil water are not directly related to the above significance to plant growth. These are discussed next.

Soil Water Constants

For a particular soil, certain soil water proportions are defined which dictate whether the water is available or not for plant growth. These are called the soil water constants, which are described below.

- Saturation capacity: this is the total water content of the soil when all the pores of the soil are filled with water. It is also termed as the maximum water holding capacity of the soil. At saturation capacity, the soil moisture tension is almost equal to zero.

- Field capacity: this is the water retained by an initially saturated soil against the force of gravity. Hence, as the gravitational water gets drained off from the soil, it is said to reach the field capacity. At field capacity, the macro-pores of the soil are drained off, but water is retained in the micropores. Though the soil moisture tension at field capacity varies from soil to soil, it is normally between 1/10 (for clayey soils) to 1/3 (for sandy soils) atmospheres.

- Permanent wilting point: plant roots are able to extract water from a soil matrix, which is saturated up to field capacity. However, as the water extraction proceeds, the moisture content diminishes and the negative (gauge) pressure increases. At one point, the plant cannot extract any further water and thus wilts.

Two stages of wilting points are recognized and they are:

- Temporary wilting point: this denotes the soil water content at which the plant wilts at day time, but recovers during right or when water is added to the soil.

- Ultimate wilting point: at such a soil water content, the plant wilts and fails to regain life even after addition of water to soil.

It must be noted that the above water contents are expressed as percentage of water held in the soil pores, compared to a fully saturated soil. Figure 2 explains graphically, the various soil constants; the full pie represents the volume of voids in soil.

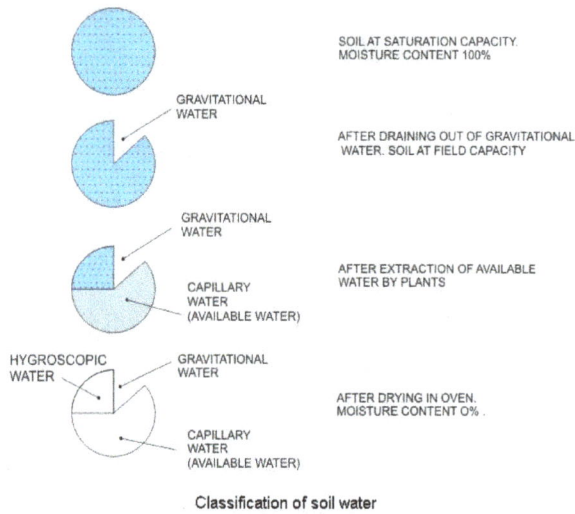

Classification of soil water

The available water for plants is defined as the difference in moisture content of the soil between field capacity and permanent wilting point.

Field capacity and Permanent wilting point: Although the pie diagrams in the above figure demonstrate the drying up of saturated soil pores, all the soil constants are expressed as a percentage by weight of the moisture available at that point compared to the weight of the dried soil sand sample.

Soil Water Constants Expressed in Depth Units

The soil water constants are mentioned as being expressed as weight percentages of the moisture content (that is amount of water) held by the water at a certain state with respect to the weight of the dried soil sample. The same may also be expressed as volume of water stored in the root zone of a field per unit area. This would consequently express the soil water constants as units of depths. The conversion from one form to the other is presented below:

Assume the following:

- Root zone depth = D (m)

- Specific weight of soil = γ_s (kg/m³)

- Specific weight of water = γ_w (kg/m³)

- Area of plot considered = 1m × 1m

Hence, the weight of soil per unit area would be: $\gamma_s \times 1 \times D$ (kg)

The weight of water held by the soil per unit area would be equal to: $\gamma \times 1 \times d$

Where d is equivalent depth of water that is actually distributed within the soil pores. Hence the following constants may be expressed as:

$$\text{Field Capacity} = \frac{\text{Weight of water held by soil per unit area}}{\text{Weight of soil per unit area}}$$

$$= \frac{\gamma_w * 1 * d}{\gamma_s * 1 * D} \qquad (1)$$

Thus, depth of water (d_{Fc}) held by soil at field capacity (FC)

$$= \frac{\gamma_w}{\gamma_s} * D * FC \qquad (2)$$

Similarly, depth of water (d_{wp}) held by soil at permanent wilting point (PWP)

$$= \frac{\gamma_s}{\gamma_w} * D * PWP \qquad (3)$$

Hence, depth of water (d_{Aw}) available to plants

$$= \frac{\gamma_s}{\gamma_w} * D * [FC - PWP] \qquad (4)$$

Therefore, the depth of water available to plants per meter depth of soil

$$= \frac{\gamma_s}{\gamma_w} [FC - PWP] \qquad (5)$$

It may be noted that plants cannot extract the full available water with the same efficiency. About 75 percent of the amount is rather easily extracted, and it is called the readily available water. The available water holding capacity for a few typical soil types are given as in the following table:

Soil Texture	Field Capacity (FC) percent	Permanent Wilting Point (PWP) percent	Bulk Density(γ_s) Kg/m³	Available water per meter depth of soil profile(m)
Sandy	5 to 10	2 to 6	1500 to 1800	0.05 to 0.1
Sandy loam	10 to 18	4 to 10	1400 to 1600	0.09 to 0.16
Loam	18 to 25	8 to 14	1300 to 1500	0.14 to 0.22

| Clay loam | 24 to 32 | 11 to 16 | 1300 to 1400 | 0.17 to 0.29 |
| Clay | 32 to 40 | 15 to 22 | 1200 to 1400 | 0.20 to 0.21 |

Water Absorption by Plants

Water is absorbed mostly through the roots of plants, though an insignificant absorption is also done through the leaves. Plants normally have a higher concentration of roots close to the soil surface and the density decreases with depth as shown in the given figure.

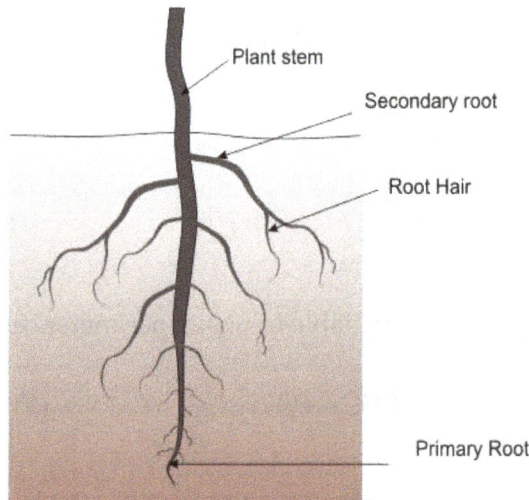

Typical root density variation of a plant with depth.

In a normal soil with good aeration, a greater portion of the roots of most plants remain within 0.45m to 0.60m of surface soil layers and most of the water needs of plants are met from this zone. As the available water from this zone decreases, plants extract more water from lower depths. When the water content of the upper soil layers reach wilting point, all the water needs of plants are met from lower layers. Since there exists few roots in lower layers, the water extract from lower layers may not be adequate to prevent wilting, although sufficient water may be available there.

When the top layers of the root zone are kept moist by frequent application of water through irrigation, plants extract most of the water (about 40 percent) from the upper quarter of their root zone. In the lower quarter of root zone the water extracted by the plant meets about 30 percent of its water needs. Further below, the third quarter of the root zone extracts about 20 percent and the lowermost quarter of root zone extracts the remaining about 10 percent of the plants water. It may be noted that the water extracted from the soil by the roots of a plant moves upwards and essentially is lost to the atmosphere as water vapours mainly through the leaves. This process, called transpiration, results in losing almost 95percent of water sucked up. Only about 5percent of water pumped up by the root system is used by the plant for metabolic purpose and increasing the plant body weight.

Importance of Water in Plant Growth

During the life cycle of a plant water, among other essential elements like air and fertilizers, plays a vital role, some of the important ones being:

- Water maintains the turgidity of the plant cells, thus keeping the plant erect. Water accounts for the largest part of the body weight of an actively growing plant and it constitutes 85 to 90 percent of the body weight of young plants and 20 to 50 percent of older or mature plants.

- Water provides both oxygen and hydrogen required for carbohydrate synthesis during the photosynthesis process.

- Water acts as a solvent of plant nutrients and helps in the uptake of nutrients from soil.

- Food manufactured in the green parts of a plant gets distributed throughout the plant body as a solution in water.

- Transpiration is a vital process in plants and does so at a maximum rate (called the potential evapo transpiration rate) when water is available in adequate amount. If soil moisture is not sufficient, then the transpiration rate is curtailed, seriously affecting plant growth and yield.

- Leaves get heated up with solar radiation and plants help to dissipate the heat by transpiration, which itself uses plant water.

Irrigation Water Quality

In irrigation agriculture, the quality of water used for irrigation should receive adequate attention. Irrigation water, regardless of its source, always contains some soluble salts in it. Apart from the total concentration of the dissolved salts, the concentration of some of the individual salts, and especially those which are most harmful to crops, is important in determining the suitability of water for irrigation. The constituents usually determined by analyzing irrigation water are the electrical conductivity for the total dissolved salts, soluble sodium percentage, sodium absorption ratio, boron content, PH, cations such as calcium, magnesium, sodium, potassium and anions such as carbonates, bicarbonates, sulphates, chlorides and nitrates.

Water from rivers which flow over salt effected areas or in the deltaic regions has a greater concentration of salts sometimes as high as 7500 ppm or even more. The quality of tank or lake water depends mainly on the soil salinity in the water shed areas and the aridity of the region. The quality of ground water resources, that is, from shallow or deep wells, is generally poor under the situations of

- high aridity

- high water table and water logged conditions

- in the vicinity of sea water

on the basis of suitability of water for irrigation, the water may be classified under three categories, which are shown in the following table:

Class	Electric al Conduc tivity (mi-cro- ohm/cm)	Total Dis-solv ed Sol-ids (ppm)	Exchangea ble sodium (percentag e)	Chlorid e (ppm)	Sulphat es (ppm)	Boron (ppm)	Remarks
I	0-1000	0-700	0-60	0-142	0-192	0-0.5	Excellent to good for irrigation
II	1000-3000	700- 2000	60-75	142- 355	192-480	0.5-2.0	Good to injuri-ous; suitable only with permeable soils and mod-erate teaching. Harmful to more sensitive crops.
III	>3000	>2000	>75	>355	>480	>2.0	Unfit for irriga-tion

Environmental Impact of Irrigation

The environmental impacts of irrigation relate to the changes in quantity and quality of soil and water as a result of irrigation and the effects on natural and social conditions in river basins and downstream of an irrigation scheme. The impacts stem from the al-tered hydrological conditions caused by the installation and operation of the irrigation scheme.

Direct Effects

An irrigation scheme draws water from groundwater, rivers, lakes or overland flow, and distributes it over an area. Hydrological, or direct, effects of doing this include reduction in downstream river flow, increased evaporation in the irrigated area, increased level in the water table as groundwater recharge in the area is increased and flow increased in the irrigated area. Likewise, irrigation has immediate effects on the provision of moisture to the atmosphere, inducing atmospheric instabilities and increasing downwind rainfall, or in other cases modifies the atmospheric circulation, delivering rain to different downwind areas. Increases or decreases in irrigation are a key area of concern in precipitationshed studies, that examine how significant modifications to the delivery of evaporation to the atmosphere can alter downwind rainfall.

Indirect Effects

Indirect effects are those that have consequences that take longer to develop and may also be longer-lasting. The indirect effects of irrigation include the following:

- Waterlogging

- Soil salination

- Ecological damage

- Socioeconomic impacts

The indirect effects of waterlogging and soil salination occur directly on the land being irrigated. The ecological and socioeconomic consequences take longer to happen but can be more far-reaching.

Some irrigation schemes use water wells for irrigation. As a result, the overall water level decreases. This may cause water mining, land/soil subsidence, and, along the coast, saltwater intrusion.

Irrigated land area worldwide occupies about 16% of the total agricultural area and the crop yield of irrigated land is roughly 40% of the total yield. In other words, irrigated land produces 2.5 times more product than non-irrigated land. This article will discuss some of the environmental and socioeconomic impacts of irrigation.

Adverse Impacts

Reduced River Flow

The reduced downstream river flow may cause:

- reduced downstream flooding

- disappearance of ecologically and economically important wetlands or flood forests

- reduced availability of industrial, municipal, household, and drinking water

- reduced shipping routes. Water withdrawal poses a serious threat to the Ganges. In India, barrages control all of the tributaries to the Ganges and divert roughly 60 percent of river flow to irrigation

- reduced fishing opportunities. The Indus River in Pakistan faces scarcity due to over-extraction of water for agriculture. The Indus is inhabited by 25 amphibian species and 147 fish species of which 22 are found nowhere else in the world. It harbors the endangered Indus River dolphin, one of the world's rarest mammals. Fish populations, the main source of protein and overall life support systems for many communities, are also being threatened

- reduced discharge into the sea, which may have various consequences like coastal erosion (e.g. in Ghana) and salt water intrusion in delta's and estuaries. Current water withdrawal from the river Nile for irrigation is so high that, despite its size, in dry periods the river does not reach the sea. The Aral Sea has suffered an "environmental catastrophe" due to the interception of river water for irrigation purposes.

Increased Groundwater Recharge, Waterlogging, Soil Salinity

Looking over the shoulder of a Peruvian farmer in the Huarmey delta at waterlogged and salinised irrigated land with a poor crop stand.
This illustrates an environmental impact of upstream irrigation developments causing an increased flow of groundwater to this lower-lying area, leading to adverse conditions.

Increased groundwater recharge stems from the unavoidable deep percolation losses occurring in the irrigation scheme. The lower the irrigation efficiency, the higher the losses. Although fairly high irrigation efficiencies of 70% or more (i.e. losses of 30% or less) can occur with sophisticated techniques like sprinkler irrigation and drip irrigation, or by well managed surface irrigation, in practice the losses are commonly in the order of 40% to 60%. This may cause the following issues:

- rising water tables

- increased storage of groundwater that may be used for irrigation, municipal, household and drinking water by pumping from wells

- waterlogging and drainage problems in villages, agricultural lands, and along roads - with mostly negative consequences. The increased level of the water table can lead to reduced agricultural production.

- shallow water tables - a sign that the aquifer is unable to cope with the ground-water recharge stemming from the deep percolation losses

- where water tables are shallow, the irrigation applications are reduced. As a result, the soil is no longer leached and soil salinity problems develop

- stagnant water tables at the soil surface are known to increase the incidence of water-borne diseases like malaria, filariasis, yellow fever, dengue, and schisto-somiasis (Bilharzia) in many areas. Health costs, appraisals of health impacts and mitigation measures are rarely part of irrigation projects, if at all.

- to mitigate the adverse effects of shallow water tables and soil salinization, some form of watertable control, soil salinity control, drainage and drainage system is needed

- as drainage water moves through the soil profile it may dissolve nutrients (either fertilizer-based or naturally occurring) such as nitrates, leading to a buildup of those nutrients in the ground-water aquifer. High nitrate levels in drinking water can be harmful to humans, particularly infants under 6 months, where it is linked to "blue-baby syndrome".

Reduced Downstream River Water Quality

Owing to drainage of surface and groundwater in the project area, which waters may be salinized and polluted by agricultural chemicals like biocides and fertilizers, the quality of the river water below the project area can deteriorate, which makes it less fit for industrial, municipal and household use. It may lead to reduced public health. Polluted river water entering the sea may adversely affect the ecology along the sea shore.

The natural build up of sedimentation can reduce downstream river flows due to the installation of irrigation systems. Sedimentation is an essential part of the ecosystem that requires the natural flux of the river flow. This natural cycle of sediment dispersion replenishes the nutrients in the soil, that will in turn, determine the livelihood of the plants and animals that rely on the sediments carried downstream. The benefits of heavy deposits of sedimentation can be seen in large rivers like the Nile River. The sediment from the delta has built up to form a giant aquifer during flood season, and retains water in the wetlands. The wetlands that are created and sustained due to built up sediment at the basin of the river is a habitat for numerous species of birds. However, heavy sedimentation can reduce downstream river water quality and can exacerbate floods up stream. This has been known to happen in the Sanmenxia reservoir in China.

The Sanmenxia reservoir is part of a larger man-made project of hydro-electric dams called the Three Gorge Project In 1998, uncertain calculations and heavy sediment greatly affected the reservoir's ability to properly fulfill its flood-control function This also reduces the down stream river water quality. Shifting more towards mass irriga-tion installments in order to meet more socioeconomic demands is going against the natural balance of nature, and use water pragmatically- use it where it is found

Affected Downstream Water Users

Water becomes scarce for nomadic pastoralist in Baluchistan due to new irrigation developments

Downstream water users often have no legal water rights and may fall victim of the development of irrigation.

Pastoralists and nomadic tribes may find their land and water resources blocked by new irrigation developments without having a legal recourse.

Flood-recession cropping may be seriously affected by the upstream interception of river water for irrigation purposes.

- In Baluchistan, Pakistan, the development of new small-scale irrigation proj-ects depleted the water resources of nomadic tribes traveling annually between Baluchistan and Gujarat or Rajasthan, India

- After the closure of the Kainji dam, Nigeria, 50 to 70 per cent of the downstream area of flood-recession cropping was lost

Lake Manantali, 477 km², displaced 12,000 people.

Lost Land use Opportunities

Irrigation projects may reduce the fishing opportunities of the original population and the grazing opportunities for cattle. The livestock pressure on the remaining lands may increase considerably, because the ousted traditional pastoralist tribes will have to find their subsistence and existence elsewhere, overgrazing may increase, followed by serious soil erosion and the loss of natural resources. The Manatali reservoir formed by the Manantali dam in Mali intersects the migration routes of nomadic pastoralists and destroyed 43000 ha of savannah, probably leading to overgrazing and erosion elsewhere. Further, the reservoir destroyed 120 km² of forest. The depletion of groundwater aquifers, which is caused by the suppression of the seasonal flood cycle, is damaging the forests downstream of the dam.

Groundwater Mining with Wells, Land Subsidence

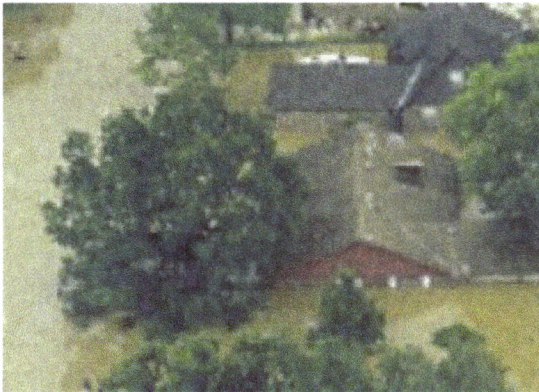

Flooding as a consequence of land subsidence

When more groundwater is pumped from wells than replenished, storage of water in the aquifer is being mined and the use of that water is no longer sustainable. As levels fail, it becomes more difficult to extract water and pumps will struggle to maintain the design flowrate and consume more may fenergy per unit of water. Eventually it may become so difficult to extract groundwater that farmers may be forced to abandon irrigated agriculture. Some notable examples include:

- The hundreds of tubewells installed in the state of Uttar Pradesh, India, with World Bank funding have operating periods of 1.4 to 4.7 hours/day, whereas they were designed to operate 16 hours/day

- In Baluchistan, Pakistan, the development of tubewell irrigation projects was at the expense of the traditional qanat or karez users

- Groundwater-related subsidence of the land due to mining of groundwater occurred in the United States at a rate of 1m for each 13m that the watertable was lowered

- Homes at Greens Bayou near Houston, Texas, where 5 to 7 feet of subsidence has occurred, were flooded during a storm in June 1989 as shown in the picture

Simulation and Prediction

The effects of irrigation on watertable, soil salinity and salinity of drainage and ground-water, and the effects of mitigative measures can be simulated and predicted using agro-hydro-salinity models like SaltMod and SahysMod

Case Studies

1. In India 2.19 million ha have been reported to suffer from waterlogging in irrigation canal commands. Also 3.47 million ha were reported to be seriously salt affected,

2. In the Indus Plains in Pakistan, more than 2 million hectares of land is water-logged. The soil of 13.6 million hectares within the Gross Command Area was surveyed, which revealed that 3.1 million hectares (23%) was saline. 23% of this was in Sindh and 13% in the Punjab. More than 3 million ha of water-logged lands have been provided with tube-wells and drains at the cost of billions of rupees, but the reclamation objectives were only partially achieved. The Asian Development Bank (ADB) states that 38% of the irrigated area is now water-logged and 14% of the surface is too saline for use

3. In the Nile delta of Egypt, drainage is being installed in millions of hectares to combat the water-logging resulting from the introduction of massive perennial irrigation after completion of the High Dam at Assuan

4. In Mexico, 15% of the 3 million ha of irrigable land is salinized and 10% is waterlogged

5. In Peru some 0.3 million ha of the 1.05 million ha of irrigable land suffers from degradation.

6. Estimates indicate that roughly one-third of the irrigated land in the major irrigation countries is already badly affected by salinity or is expected to become so in the near future. Present estimates for Israel are 13% of the irrigated land, Australia 20%, China 15%, Iraq 50%, Egypt 30%. Irrigation-induced salinity occurs in large and small irrigation systems alike

7. FAO has estimated that by 1990 about 52 million ha of irrigated land will need to have improved drainage systems installed, much of it subsurface drainage to control salinity

Reduced Downstream Drainage and Groundwater Quality

- The downstream drainage water quality may deteriorate owing to leaching of salts, nutrients, herbicides and pesticides with high salinity and alkalinity.

There is threat of soils converting into saline or alkali soils. This may negatively affect the health of the population at the tail-end of the river basin and downstream of the irrigation scheme, as well as the ecological balance. The Aral Sea, for example, is seriously polluted by drainage water.

- The downstream quality of the groundwater may deteriorate in a similar way as the downstream drainage water and have similar consequences

Mitigation of Adverse Effects

Irrigation can have a variety negative impacts on ecology and socioeconomy, which may be mitigated in a number of ways. These include siting the irrigation project in a location which minimises negative impacts. The efficiency of existing projects can be improved and existing degraded croplands can be improved rather than establishing a new irrigation project Developing small-scale, individually owned irrigation systems as an alternative to large-scale, publicly owned and managed schemes. The use of sprinkler irrigation and micro-irrigation systems decreases the risk of waterlogging and erosion. Where practicable, using treated wastewater makes more water available to other users Maintaining flood flows downstream of the dams can ensure that an adequate area is flooded each year, supporting, amongst other objectives, fishery activities.

Delayed Environmental Impacts

It often takes time to accurately predict the impact that new irrigation schemes will have on the ecology and socioeconomy of a region. By the time these predictions are available, a considerable amount of time and resources may have already been expended in the implementation of that project. When that is the case, the project managers will often only change the project if the impact would be considerably more than they had originally expected.

Potential Benefits Outweigh the Potential Disadvantages

Frequently irrigation schemes are seen as extremely necessary for socioeconomic well-being especially in developing countries. One example of this can be demonstrated from a proposal for an irrigation scheme in Malawi. Here it was shown that the potential positive effects of the irrigation project that was being proposed "outweighed the potential negative impacts". It was stated that the impacts would mostly "be localized, minimal, short term occurring during the construction and operation phases of the Project". In order to help alleviate and prevent major environmental impacts, they would use techniques that minimize the potential negative impacts. As far as the region's socioeconomic well-being, there would be no "displacement and/ or resettlement envisioned during the implementation of the Project activities". The original primary purposes of the irrigation project were to reduce poverty, improve food security, create local employment, increase household income and enhance the sustainability of land use.

Due to this careful planning this project was successful both in improving the socialeconomic conditions in the region and ensuring that land and water are sustainability into the future.

Soil Salinity

Visibly salt-affected soils on rangeland in Colorado. Salts dissolved from the soil accumulate at the soil surface and are deposited on the ground and at the base of the fence post.

Soil salinity is the salt content in the soil; the process of increasing the salt content is known as salinization. Salts occur naturally within soils and water. Salination can be caused by natural processes such as mineral weathering or by the gradual withdrawal of an ocean. It can also come about through artificial processes such as irrigation.

Natural Occurrence

Salts are a natural component in soils and water. The ions responsible for salination are: Na^+, K^+, Ca^{2+}, Mg^{2+} and Cl^-.

As the Na^+ (sodium) predominates, soils can become sodic. Sodic soils present particular challenges because they tend to have very poor structure which limits or prevents water infiltration and drainage.

Over long periods of time, as soil minerals weather and release salts, these salts are flushed or leached out of the soil by drainage water in areas with sufficient precipitation. In addition to mineral weathering, salts are also deposited via dust and precipitation. In dry regions salts may accumulate, leading to naturally saline soils. This is the case, for example, in large parts of Australia. Human practices can increase the salinity of soils by the addition of salts in irrigation water. Proper irrigation management can prevent salt accumulation by providing adequate drainage water to leach added salts from the soil. Disrupting drainage patterns that provide leaching can also result in salt accumulations.

An example of this occurred in Egypt in 1970 when the Aswan High Dam was built. The change in the level of ground water before the construction had enabled soil erosion, which led to high concentration of salts in the water table. After the construction, the continuous high level of the water table led to the salination of the arable land.

Dry Land Salinity

Salinity in drylands can occur when the water table is between two and three metres from the surface of the soil. The salts from the groundwater are raised by capillary action to the surface of the soil. This occurs when groundwater is saline (which is true in many areas), and is favored by land use practices allowing more rainwater to enter the aquifer than it could accommodate. For example, the clearing of trees for agriculture is a major reason for dryland salinity in some areas, since deep rooting of trees has been replaced by shallow rooting of annual crops.

Salinity due to Irrigation

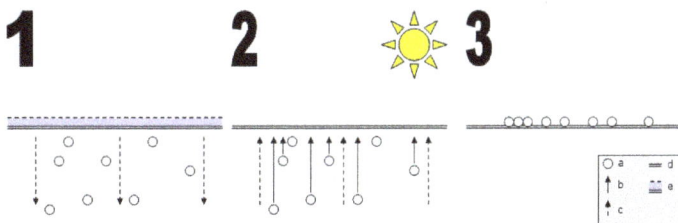

Rain or irrigation, in the absence of leaching, can bring salts to the surface by capillary action

Salinity from irrigation can occur over time wherever irrigation occurs, since almost all water (even natural rainfall) contains some dissolved salts. When the plants use the water, the salts are left behind in the soil and eventually begin to accumulate. Since soil salinity makes it more difficult for plants to absorb soil moisture, these salts must be leached out of the plant root zone by applying additional water. This water in excess of plant needs is called the leaching fraction. Salination from irrigation water is also greatly increased by poor drainage and use of saline water for irrigating agricultural crops.

Salinity in urban areas often results from the combination of irrigation and groundwater processes. Irrigation is also now common in cities (gardens and recreation areas).

Consequences of Salinity

The consequences of salinity are

- detrimental effects on plant growth and yield

- damage to infrastructure (roads, bricks, corrosion of pipes and cables)

- reduction of water quality for users, sedimentation problems

- soil erosion ultimately, when crops are too strongly affected by the amounts of salts.

Salinity is an important land degradation problem. Soil salinity can be reduced by leaching soluble salts out of soil with excess irrigation water. Soil salinity control involves watertable control and flushing in combination with tile drainage or another form of subsurface drainage. A comprehensive treatment of soil salinity is available from the United Nations Food and Agriculture Organization.

Salt Tolerance of Crops

High levels of soil salinity can be tolerated if salt-tolerant plants are grown. Sensitive crops lose their vigor already in slightly saline soils, most crops are negatively affected by (moderately) saline soils, and only salinity resistant crops thrive in severely saline soils. The University of Wyoming and the Government of Alberta report data on the salt tolerance of plants.

Field data, under farmers' conditions, in irrigated lands are scarce, especially in developing countries. However, some on farm surveys were made in Egypt, India, and Pakistan. Some examples are shown in the following gallery with crops arranged from sensitive to very tolerant.

- Graphs of crop yield and soil salinity in farmers' fields ordered by increasing salt tolerance

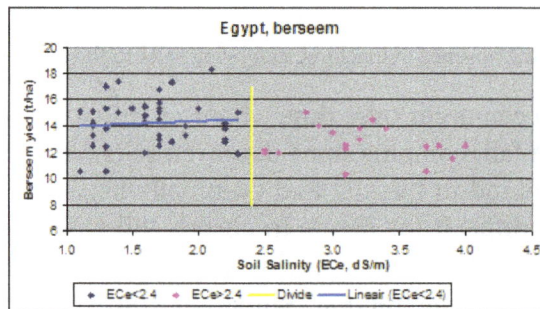

Berseem (clover), cultivated in Egypt's Nile Delta is a salt sensitive crop an tolerates an ECe value up to 2.4 dS/m whereafter yields start to decline.

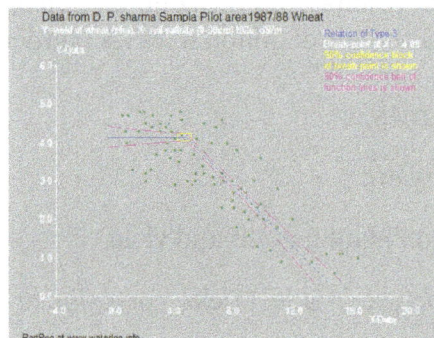

Wheat grown in Sampla, Haryana, India is slightly sensitive tolerating an ECe value of 4.9 dS/m.

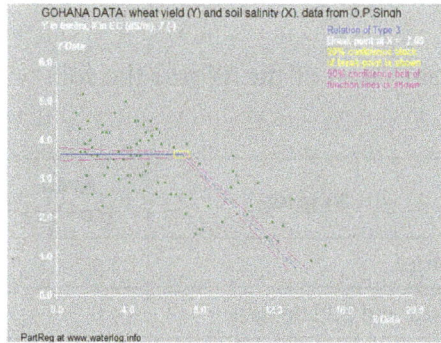

The field measurements in wheat fields in Gohana, Haryana, India, showed a higher tolerance level of ECe = 7.1 dS/m. (The Egyptian wheat, not shown here, exhibited a tolerance point of 7.8 dS/m).

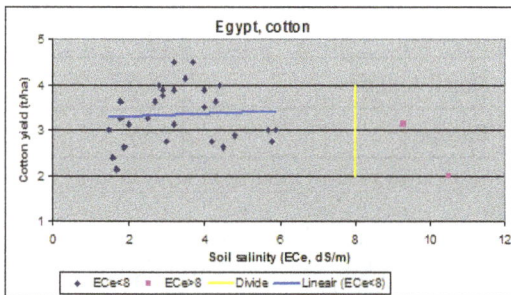

The cotton grown in the Nile Delta can be called salt just tolerant with a critical ECe value of 8.0 dS/m.

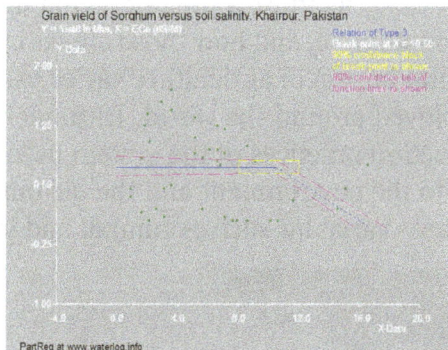

Sorghum from Khairpur, Pakistan is quite tolerant, it grows well up to ECe = 10.5 dS/m

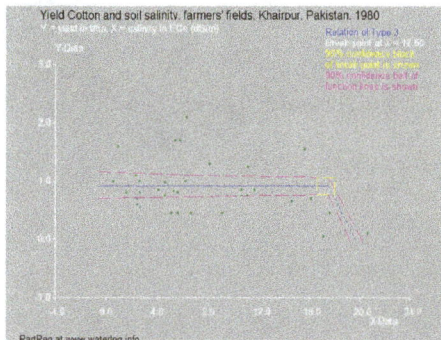

Cotton from Khairpur, Pakistan is very tolerant, it grows well up to ECe = 15.5 dS/m

Regions Affected

From the FAO/UNESCO Soil Map of the World the following salinised areas can be derived.

Region	Area (10^6 ha)
Africa	69.5
Near and Middle East	53.1
Asia and Far East	19.5
Latin America	59.4
Australia	84.7
North America	16.0
Europe	20.7

Environmental Impact of Agriculture

The environmental impact of agriculture is the effect that different farming practices have on the ecosystems around them, and how those effects can be traced back to those practices. The environmental impact of agriculture varies based on the wide variety of agricultural practices employed around the world. Ultimately, the environmental impact depends on the production practices of the system used by farmers. The connection between emissions into the environment and the farming system is indirect, as it also depends on other climate variables such as rainfall and temperature.

Water pollution in a rural stream due to runoff from farming activity in New Zealand.

There are two types of indicators of environmental impact: "means-based", which is based on the farmer's production methods, and "effect-based", which is the impact that farming methods have on the farming system or on emissions to the environment. An example of a means-based indicator would be the quality of groundwater, that is effected by the amount of nitrogen applied to the soil. An indicator reflecting the loss of nitrate to groundwater would be effect-based. The means-based evaluation looks at farmers' practices of agriculture, and the effect-based evaluation considers the actual effects of the agricultural system. For example, means-based analysis might look at pesticides and fertilization methods that farmers are using, and effect-based analysis would consider how much CO_2 is being emitted or what the Nitrogen content of the soil is.

The environmental impact of agriculture involves a variety of factors from the soil, to water, the air, animal and soil variety, people, plants, and the food itself. Some of the environmental issues that are related to agriculture are climate change, deforestation, genetic engineering, irrigation problems, pollutants, soil degradation, and waste.

Issues

Climate Change

Climate change and agriculture are interrelated processes, both of which take place on a worldwide scale. Global warming is projected to have significant impacts on conditions affecting agriculture, including temperature, precipitation and glacial run-off. These conditions determine the carrying capacity of the biosphere to produce enough food for the human population and domesticated animals. Rising carbon dioxide levels would also have effects, both detrimental and beneficial, on crop yields. Assessment of the effects of global climate changes on agriculture might help to properly anticipate and adapt farming to maximize agricultural production. Although the net impact of climate change on agricultural production is uncertain it is likely that it will shift the suitable growing zones for individual crops. Adjustment to this geographical shift will involve considerable economic costs and social impacts..

At the same time, agriculture has been shown to produce significant effects on climate change, primarily through the production and release of greenhouse gases such as carbon dioxide, methane, and nitrous oxide. In addition, agriculture that practices tillage, fertilization, and pesticide application also releases ammonia, nitrate, phosphorus, and many other pesticides that affect air, water, and soil quality, as well as biodiversity. Agriculture also alters the Earth's land cover, which can change its ability to absorb or reflect heat and light, thus contributing to radiative forcing. Land use change such as deforestation and desertification, together with use of fossil fuels, are the major anthropogenic sources of carbon dioxide; agriculture itself is the major contributor to increasing methane and nitrous oxide concentrations in earth's atmosphere.

Deforestation

Deforestation is clearing the Earth's forests on a large scale worldwide and resulting in many land damages. One of the causes of deforestation is to clear land for pasture or crops. According to British environmentalist Norman Myers, 5% of deforestation is due to cattle ranching, 19% due to over-heavy logging, 22% due to the growing sector of palm oil plantations, and 54% due to slash-and-burn farming.

Deforestation causes the loss of habitat for millions of species, and is also a driver of climate change. Trees act as a carbon sink: that is, they absorb carbon dioxide, an unwanted greenhouse gas, out of the atmosphere. Removing trees releases carbon dioxide into the atmosphere and leaves behind fewer trees to absorb the increasing amount of carbon dioxide in the air. In this way, deforestation exacerbates climate change. When trees are removed from forests, the soils tend to dry out because there is no longer shade, and there are not enough trees to assist in the water cycle by returning water vapor back to the environment. With no trees, landscapes that were once forests can potentially become barren deserts. The removal of trees also causes extreme fluctuations in temperature.

In 2000 the United Nations Food and Agriculture Organization (FAO) found that "the role of population dynamics in a local setting may vary from decisive to negligible," and that deforestation can result from "a combination of population pressure and stagnating economic, social and technological conditions."

Genetic Engineering

Genetically engineered crops are herbicide-tolerant, and their overuse has created herbicide resistant "super weeds", which may ultimately increase the use of herbicides. Seed contamination is another problem of genetic engineering; it can occur from wind or bee pollination that is blown from genetically-engineered crops to normal crops. About 50% of corn and soybean samples and more than 80% of canola samples were found to be contaminated by Monsanto's (genetic engineering company) genes. This accidental contamination can cause organic farmers to lose a lot of money because they need to recall their products. There are various cases of this such as in the corn and alfalfa industry.

Irrigation

Irrigation can lead to a number of problems:

Among some of these problems is the depletion of underground aquifers through over-drafting. Soil can be over-irrigated because of poor distribution uniformity or management wastes water, chemicals, and may lead to water pollution. Over-irrigation can cause deep drainage from rising water tables that can lead to problems of irrigation salinity requiring watertable control by some form of subsurface land drainage. However, if the soil is under irrigated, it gives poor soil salinity control which leads to increased soil salinity with consequent buildup of toxic salts on soil surface in areas with high evaporation. This requires either leaching to remove these salts and a method of drainage to carry the salts away. Irrigation with saline or high-sodium water may damage soil structure owing to the formation of alkaline soil.

Pollutants

Synthetic pesticides such as 'Malathon, 'Rogor', 'Kelthane' and 'confidor' are the most widespread method of controlling pests in agriculture. Pesticides can leach through the soil and enter the groundwater, as well as linger in food products and result in death in humans. Pesticides can also kill non-target plants, birds, fish and other wildlife. A wide range of agricultural chemicals are used and some become pollutants through use, misuse, or ignorance. Pollutants from agriculture have a huge effect on water quality. Agricultural nonpoint source (NPS) solution impacts lakes, rivers, wetlands, estuaries, and groundwater. Agricultural NPS can be caused by poorly managed animal feeding operations, overgrazing, plowing, fertilizer, and improper, excessive, or badly timed use of Pesticides. Pollutants from farming include sediments, nutrients, pathogens, pesticides, metals, and salts.

Listed below are additional and specific problems that may arise with the release of pollutants from agriculture.

- Pesticide drift

  o   soil contamination

  o   air pollution *spray drift*

- Pesticides, especially those based on organochloride

- Pesticide residue in foods

- Pesticide toxicity to bees

  o   List of crop plants pollinated by bees

  o   Pollination management

- Bioremediation

Soil Degradation

Soil degradation is the decline in soil quality that can be a result of many factors, especially from agriculture. Soils hold the majority of the world's biodiversity, and healthy soils are essential for food production and an adequate water supply. Common attributes of soil degradation can be salting, waterlogging, compaction, pesticide contamination, decline in soil structure quality, loss of fertility, changes in soil acidity, alkalinity, salinity, and erosion. Soil degradation also has a huge impact on biological degradation, which affects the microbial community of the soil and can alter nutrient cycling, pest and disease control, and chemical transformation properties of the soil.

- soil contamination
  - o sedimentation

Waste

Plasticulture is the use of plastic mulch in agriculture. Farmers use plastic sheets as mulch to cover 50-70% of the soil and allows them to use drip irrigation systems to have better control over soil nutrients and moisture. Rain is not required in this system, and farms that use plasticulture are built to encourage the fastest runoff of rain. The use of pesticides with plasticulture allows pesticides to be transported easier in the surface runoff towards wetlands or tidal creeks. The runoff from pesticides and chemicals in the plastic can cause serious deformations and death in shellfish as the runoff carries the chemicals towards the oceans.

In addition to the increased runoff that results from plasticulture, there is also the problem of the increased amount of waste form the plastic mulch itself. The use of plastic mulch for vegetables, strawberries, and other row and orchard crops exceeds 110 million pounds annually in the United States. Most plastic ends up in the landfill, although there are other disposal options such as disking mulches into the soil, on-site burying, on-site storage, reuse, recycling, and incineration. The incineration and recycling options are complicated by the variety of the types of plastics that are used and by the geographic dispersal of the plastics. Plastics also contain stabilizers and dyes as well as heavy metals, which limits the amount of products that can be recycled. Research is continually being conducted on creating biodegradable or photodegradable mulches. While there has been minor success with this, there is also the problem of how long the plastic takes to degrade, as many biodegradable products take a long time to break down.

Issues by Region

The environmental impact of agriculture can vary depending on the region as well as the type of agriculture production method that is being used. Listed below are some specific environmental issues in a various different regions around the world.

- Hedgerow removal in the United Kingdom.

- Soil salinisation, especially in Australia.

- Phosphate mining in Nauru

- Methane emissions from livestock in New Zealand.

- Environmentalists attribute the hypoxic zone in the Gulf of Mexico as being encouraged by nitrogen fertilization of the algae bloom.

Sustainable Agriculture

Sustainable agriculture is the idea that agriculture should occur in a way such that we can continue to produce what is necessary without infringing on the ability for future generations to do the same.

The exponential population increase in recent decades has increased the practice of agricultural land conversion to meet demand for food which in turn has increased the effects on the environment. The global population is still increasing and will eventually stabilise, as some critics doubt that food production, due to lower yields from global warming, can support the global population.

Agriculture can have negative effects on biodiversity as well. Organic farming is a multifaceted sustainable agriculture set of practices that can have a lower impact on the environment at the small scale. However in most cases organic farming results in lower yields in terms of production per unit area and per unit of irrigation water. Therefore, widespread adoption of organic agriculture will require additional land to be cleared and water resources extracted to meet the same level of production. A European meta-analysis found that organic farms tended to have higher soil organic matter content and lower nutrient losses (nitrogen leaching, nitrous oxide emissions and ammonia emissions) per unit of field area but higher ammonia emissions, nitrogen leaching and nitrous oxide emissions per product unit. It is believed by many that conventional farming systems cause less rich biodiversity than organic systems. Organic farming has shown to have on average 30% higher species richness than conventional farming. Organic systems on average also have 50% more organisms. This data has some issues because there were several results that showed a negative effect on these things when in an organic farming system. The opposition to organic agriculture believes that these negatives are an issue with the organic farming system. What began as a small scale, environmentally conscious has now become just as industrialized as conventional agriculture. This industrialization can lead to the issues shown above such as climate change, and deforestation.

Conservation Tillage

Conservation tillage is a alternative tillage method for farming which is more sustainable for the soil and surrounding ecosystem. This is done by allowing the residue of the previous harvest's crops to remain in the soil before tilling for the next crop. Conservation tillage has shown to improve many things such as soil moisture retention, and reduce erosion. Some disadvantages are the fact that more expensive equiptment is needed for this process, more pesticides will need to be used, and the positive effects take a long time to be visible. The barriers of instantiating a conservation tillage poliy are that farmers are reluctant to change their methods, and would protest a more expensive, and time consuming method of tillage than the conventional one they are used to.

Other specific methods include: permaculture; and biodynamic agriculture which incorporates a spiritual element.

- Category: Sustainable agriculture

- Biological pest control

References

- O. A. Tuinenburg et al., The fate of evaporated water from the Ganges basin, Journal of Geophysical Research: Atmospheres, Volume 117, Issue D1, 16 January 2012

- N.T. Singh, 2005. Irrigation and soil salinity in the Indian subcontinent: past and present. Lehigh University Press. ISBN 0-934223-78-5, ISBN 978-0-934223-78-2, 404

- Kidd, Greg (1999–2000). "Pesticides and Plastic Mulch Threaten the Health of Maryland and Virginia East Shore Waters" (PDF). Pesticides and You. 19 (4): 22–23. Retrieved 23 April 2015

- Drainage Manual: A Guide to Integrating Plant, Soil, and Water Relationships for Drainage of Irrigated Lands, Interior Dept., Bureau of Reclamation, 1993, ISBN 0-16-061623-9

- "Irrigation potential in Africa: A basin approach". Natural Resources Management and Environment Department. Retrieved 13 March 2014

4

Principle Sources of Water Pollution

The presence of substances in water that is harmful to biotic and abiotic elements is called water pollution. Water pollution can be classified into surface water pollution and groundwater pollution. While the former concentrates on water pollution at the surface level, the latter studies the pollution keeping groundwater in focus. The chapter closely examines the principle sources of surface water pollution to provide an extensive understanding of the subject.

Water Pollution

Raw sewage and industrial waste in the New River as it passes from Mexicali to Calexico, California

Water pollution is the contamination of water bodies (e.g. lakes, rivers, oceans, aquifers and groundwater). This form of environmental degradation occurs when pollutants are directly or indirectly discharged into water bodies without adequate treatment to remove harmful compounds.

Water pollution affects the entire biosphere – plants and organisms living in these bodies of water. In almost all cases the effect is damaging not only to individual species and population, but also to the natural biological communities.

Water pollution is a major global problem which requires ongoing evaluation and revision of water resource policy at all levels (international down to individual aquifers and wells). It has been suggested that water pollution is the leading worldwide cause of deaths and diseases, and that it accounts for the deaths of more than 14,000 people daily. An estimated 580 people in India die of water pollution related illness every day. About 90 percent of the water in the cities of China is polluted. As of 2007, half a billion Chinese had no access to safe drinking water. In addition to the acute problems of water pollution in developing countries, developed countries also continue to struggle with pollution problems. For example, in the most recent national report on water quality in the United States, 44 percent of assessed stream miles, 64 percent of assessed lake acres, and 30 percent of assessed bays and estuarine square miles were classified as polluted. The head of China's national development agency said in 2007 that one quarter the length of China's seven main rivers were so poisoned the water harmed the skin.

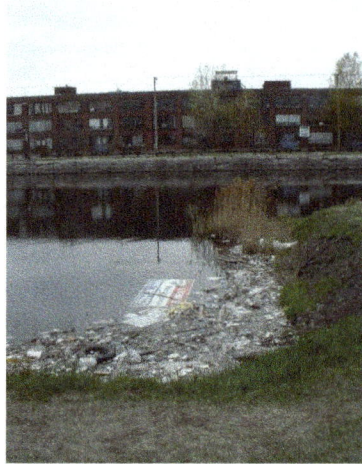

Pollution in the Lachine Canal, Canada

Water is typically referred to as polluted when it is impaired by anthropogenic contaminants and either does not support a human use, such as drinking water, or undergoes a marked shift in its ability to support its constituent biotic communities, such as fish. Natural phenomena such as volcanoes, algae blooms, storms, and earthquakes also cause major changes in water quality and the ecological status of water.

Categories

Although interrelated, surface water and groundwater have often been studied and managed as separate resources. Surface water seeps through the soil and becomes groundwater. Conversely, groundwater can also feed surface water sources. Sources of surface water pollution are generally grouped into two categories based on their origin.

Point Sources

Point source water pollution refers to contaminants that enter a waterway from a single, identifiable source, such as a pipe or ditch. Examples of sources in this category include discharges from a sewage treatment plant, a factory, or a city storm drain. The U.S. Clean Water Act (CWA) defines point source for regulatory enforcement purposes. The CWA definition of point source was amended in 1987 to include municipal storm sewer systems, as well as industrial storm water, such as from construction sites.

Point source pollution at a shipyard in Rio de Janeiro, Brazil.

Non-point Sources

Nonpoint source pollution refers to diffuse contamination that does not originate from a single discrete source. NPS pollution is often the cumulative effect of small amounts of contaminants gathered from a large area. A common example is the leaching out of nitrogen compounds from fertilized agricultural lands. Nutrient run-off in storm water from "sheet flow" over an agricultural field or a forest are also cited as examples of NPS pollution.

Blue drain and yellow fish symbol used by the UK Environment Agency to raise awareness of the ecological impacts of contaminating surface drainage

Contaminated storm water washed off of parking lots, roads and highways, called urban runoff, is sometimes included under the category of NPS pollution. However, because this runoff is typically channeled into storm drain systems and discharged through pipes to local surface waters, it becomes a point source.

Groundwater Pollution

Interactions between groundwater and surface water are complex. Consequently, groundwater pollution, also referred to as groundwater contamination, is not as easily classified as surface water pollution. By its very nature, groundwater aquifers are susceptible to contamination from sources that may not directly affect surface water bodies, and the distinction of point vs. non-point source may be irrelevant. A spill or ongoing release of chemical or radionuclide contaminants into soil (located away from a surface water body) may not create point or non-point source pollution but can contaminate the aquifer below, creating a toxic plume. The movement of the plume, called a plume front, may be analyzed through a hydrological transport model or groundwater model. Analysis of groundwater contamination may focus on soil characteristics and site geology, hydrogeology, hydrology, and the nature of the contaminants.

Causes

The specific contaminants leading to pollution in water include a wide spectrum of chemicals, pathogens, and physical changes such as elevated temperature and discoloration. While many of the chemicals and substances that are regulated may be naturally occurring (calcium, sodium, iron, manganese, etc.) the concentration is often the key in determining what is a natural component of water and what is a contaminant. High concentrations of naturally occurring substances can have negative impacts on aquatic flora and fauna.

Oxygen-depleting substances may be natural materials such as plant matter (e.g. leaves and grass) as well as man-made chemicals. Other natural and anthropogenic substances may cause turbidity (cloudiness) which blocks light and disrupts plant growth, and clogs the gills of some fish species.

Many of the chemical substances are toxic. Pathogens can produce waterborne diseases in either human or animal hosts. Alteration of water's physical chemistry includes acidity (change in pH), electrical conductivity, temperature, and eutrophication. Eutrophication is an increase in the concentration of chemical nutrients in an ecosystem to an extent that increases the primary productivity of the ecosystem. Depending on the degree of eutrophication, subsequent negative environmental effects such as anoxia (oxygen depletion) and severe reductions in water quality may occur, affecting fish and other animal populations.

Pathogens

Disease-causing microorganisms are referred to as pathogens. Although the vast majority of bacteria are either harmless or beneficial, a few pathogenic bacteria can cause disease. Coliform bacteria, which are not an actual cause of disease, are commonly used as a bacterial indicator of water pollution. Other microorganisms sometimes found in surface waters that have caused human health problems include:

- *Burkholderia pseudomallei*

- *Cryptosporidium parvum*

- *Giardia lamblia*

- *Salmonella*

- *Norovirus* and other viruses

- *Parasitic worms including the Schistosoma type*

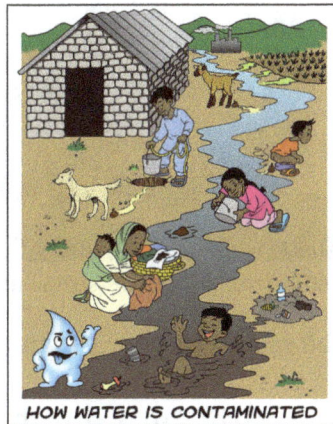

Poster to teach people in South Asia about human activities leading to the pollution of water sources

A manhole cover unable to contain a sanitary sewer overflow.

Fecal sludge collected from pit latrines is dumped into a river at the Korogocho slum in Nairobi, Kenya.

High levels of pathogens may result from on-site sanitation systems (septic tanks, pit latrines) or inadequately treated sewage discharges. This can be caused by a sewage plant designed with less than secondary treatment (more typical in less-developed countries). In developed countries, older cities with aging infrastructure may have leaky sewage collection systems (pipes, pumps, valves), which can cause sanitary sewer overflows. Some cities also have combined sewers, which may discharge untreated sewage during rain storms.

Muddy river polluted by sediment.

Pathogen discharges may also be caused by poorly managed livestock operations.

Organic, Inorganic and Macroscopic Contaminants

Contaminants may include organic and inorganic substances.

A garbage collection boom in an urban-area stream in Auckland, New Zealand.

Organic water pollutants include:

- Detergents

- Disinfection by-products found in chemically disinfected drinking water, such as chloroform

- Food processing waste, which can include oxygen-demanding substances, fats and grease

- Insecticides and herbicides, a huge range of organohalides and other chemical compounds

- Petroleum hydrocarbons, including fuels (gasoline, diesel fuel, jet fuels, and fuel oil) and lubricants (motor oil), and fuel combustion byproducts, from storm water runoff

- Volatile organic compounds, such as industrial solvents, from improper storage.

- Chlorinated solvents, which are dense non-aqueous phase liquids, may fall to the bottom of reservoirs, since they don't mix well with water and are denser.

    o Polychlorinated biphenyl (PCBs)

    o Trichloroethylene

- Perchlorate

- Various chemical compounds found in personal hygiene and cosmetic products

- Drug pollution involving pharmaceutical drugs and their metabolites

Inorganic water pollutants include:

- Acidity caused by industrial discharges (especially sulfur dioxide from power plants)

- Ammonia from food processing waste

- Chemical waste as industrial by-products

- Fertilizers containing nutrients--nitrates and phosphates—which are found in storm water runoff from agriculture, as well as commercial and residential use

- Heavy metals from motor vehicles (via urban storm water runoff) and acid mine drainage

- Secretion of creosote preservative into the aquatic ecosystem

- Silt (sediment) in runoff from construction sites, logging, slash and burn practices or land clearing sites.

Macroscopic pollution – large visible items polluting the water – may be termed "floatables" in an urban storm water context, or marine debris when found on the open seas, and can include such items as:

- Trash or garbage (e.g. paper, plastic, or food waste) discarded by people on the ground, along with accidental or intentional dumping of rubbish, that are washed by rainfall into storm drains and eventually discharged into surface waters

- Nurdles, small ubiquitous waterborne plastic pellets

- Shipwrecks, large derelict ships.

The Brayton Point Power Station in Massachusetts discharges heated water to Mount Hope Bay.

Thermal Pollution

Thermal pollution is the rise or fall in the temperature of a natural body of water caused by human influence. Thermal pollution, unlike chemical pollution, results in a change in the physical properties of water. A common cause of thermal pollution is the use of water as a coolant by power plants and industrial manufacturers. Elevated water temperatures decrease oxygen levels, which can kill fish and alter food chain composition, reduce species biodiversity, and foster invasion by new thermophilic species. Urban runoff may also elevate temperature in surface waters.

Thermal pollution can also be caused by the release of very cold water from the base of reservoirs into warmer rivers.

Transport and Chemical Reactions of Water Pollutants

Most water pollutants are eventually carried by rivers into the oceans. In some areas of the world the influence can be traced one hundred miles from the mouth by studies using hydrology transport models. Advanced computer models such as SWMM or the DSSAM Model have been used in many locations worldwide to examine the fate of pollutants in aquatic systems. Indicator filter-feeding species such as copepods have also been used to study pollutant fates in the New York Bight, for example. The highest toxin loads are not directly at the mouth of the Hudson River, but 100 km (62 mi) south, since several days are required for incorporation into planktonic tissue. The Hudson discharge flows south along the coast due to the coriolis force. Further south are areas of oxygen depletion caused by chemicals using up oxygen and by algae blooms, caused by excess nutrients from algal cell death and decomposition. Fish and shellfish kills have been reported, because toxins climb the food chain after small fish consume copepods, then large fish eat smaller fish, etc. Each successive step up the food chain causes a cumulative concentration of pollutants such as heavy metals (e.g. mercury) and persistent organic pollutants such as DDT. This is known as bio-magnification, which is occasionally used interchangeably with bio-accumulation.

A polluted river draining an abandoned copper mine on Anglesey

Large gyres (vortexes) in the oceans trap floating plastic debris. The North Pacific Gyre, for example, has collected the so-called "Great Pacific Garbage Patch", which is now estimated to be one hundred times the size of Texas. Plastic debris can absorb toxic chemicals from ocean pollution, potentially poisoning any creature that eats it. Many of these long-lasting pieces wind up in the stomachs of marine birds and animals. This results in obstruction of digestive pathways, which leads to reduced appetite or even starvation.

Many chemicals undergo reactive decay or chemical change, especially over long periods of time in groundwater reservoirs. A noteworthy class of such chemicals is the chlorinated hydrocarbons such as trichloroethylene (used in industrial metal degreasing and electronics manufacturing) and tetrachloroethylene used in the dry cleaning industry. Both of these chemicals, which are carcinogens themselves, undergo partial decomposition reactions, leading to new hazardous chemicals (including dichloroethylene and vinyl chloride).

Groundwater pollution is much more difficult to abate than surface pollution because groundwater can move great distances through unseen aquifers. Non-porous aquifers such as clays partially purify water of bacteria by simple filtration (adsorption and absorption), dilution, and, in some cases, chemical reactions and biological activity; however, in some cases, the pollutants merely transform to soil contaminants. Groundwater that moves through open fractures and caverns is not filtered and can be transported as easily as surface water. In fact, this can be aggravated by the human tendency to use natural sinkholes as dumps in areas of karst topography.

There are a variety of secondary effects stemming not from the original pollutant, but a derivative condition. An example is silt-bearing surface runoff, which can inhibit the penetration of sunlight through the water column, hampering photosynthesis in aquatic plants.

Measurement

Environmental scientists preparing water autosamplers.

Water pollution may be analyzed through several broad categories of methods: physical, chemical and biological. Most involve collection of samples, followed by specialized analytical tests. Some methods may be conducted *in situ*, without sampling, such as temperature. Government agencies and research organizations have published standardized, validated analytical test methods to facilitate the comparability of results from disparate testing events.

Sampling

Sampling of water for physical or chemical testing can be done by several methods, depending on the accuracy needed and the characteristics of the contaminant. Many contamination events are sharply restricted in time, most commonly in association with rain events. For this reason "grab" samples are often inadequate for fully quantifying contaminant levels. Scientists gathering this type of data often employ auto-sampler devices that pump increments of water at either time or discharge intervals.

Sampling for biological testing involves collection of plants and/or animals from the surface water body. Depending on the type of assessment, the organisms may be identified for biosurveys (population counts) and returned to the water body, or they may be dissected for bioassays to determine toxicity.

Physical Testing

Common physical tests of water include temperature, solids concentrations (e.g., total suspended solids (TSS)) and turbidity.

Chemical Testing

Water samples may be examined using the principles of analytical chemistry. Many published test methods are available for both organic and inorganic compounds. Frequently used methods include pH, biochemical oxygen demand (BOD), chemical oxygen demand (COD), nutrients (nitrate and phosphorus compounds), metals (including copper, zinc, cadmium, lead and mercury), oil and grease, total petroleum hydrocarbons (TPH), and pesticides.

Biological Testing

Biological testing involves the use of plant, animal, and/or microbial indicators to monitor the health of an aquatic ecosystem. They are any biological species or group of species whose function, population, or status can reveal what degree of ecosystem or environmental integrity is present. One example of a group of bio-indicators are the copepods and other small water crustaceans that are present in many water bodies. Such organisms can be monitored for changes (biochemical, physiological, or behavioral) that may indicate a problem within their ecosystem.

Control of Pollution

Decisions on the type and degree of treatment and control of wastes, and the disposal and use of adequately treated wastewater, must be based on a consideration all the technical factors of each drainage basin, in order to prevent any further contamination or harm to the environment.

Sewage Treatment

Deer Island Wastewater Treatment Plant serving Boston, Massachusetts and vicinity.

In urban areas of developed countries, domestic sewage is typically treated by centralized sewage treatment plants. Well-designed and operated systems (i.e., secondary treatment or better) can remove 90 percent or more of the pollutant load in sewage. Some plants have additional systems to remove nutrients and pathogens.

Cities with sanitary sewer overflows or combined sewer overflows employ one or more engineering approaches to reduce discharges of untreated sewage, including:

- utilizing a green infrastructure approach to improve storm water management capacity throughout the system, and reduce the hydraulic overloading of the treatment plant

- repair and replacement of leaking and malfunctioning equipment

- increasing overall hydraulic capacity of the sewage collection system (often a very expensive option).

A household or business not served by a municipal treatment plant may have an individual septic tank, which pre-treats the wastewater on site and infiltrates it into the soil.

Industrial Wastewater Treatment

Dissolved air flotation system for treating industrial wastewater.

Some industrial facilities generate ordinary domestic sewage that can be treated by municipal facilities. Industries that generate wastewater with high concentrations of conventional pollutants (e.g. oil and grease), toxic pollutants (e.g. heavy metals, volatile organic compounds) or other non-conventional pollutants such as ammonia, need specialized treatment systems. Some of these facilities can install a pre-treatment system to remove the toxic components, and then send the partially treated wastewater to the municipal system. Industries generating large volumes of wastewater typically operate their own complete on-site treatment systems. Some industries have been successful at redesigning their manufacturing processes to reduce or eliminate pollutants, through a process called pollution prevention.

Heated water generated by power plants or manufacturing plants may be controlled with:

- cooling ponds, man-made bodies of water designed for cooling by evaporation, convection, and radiation

- cooling towers, which transfer waste heat to the atmosphere through evaporation and/or heat transfer

- cogeneration, a process where waste heat is recycled for domestic and/or industrial heating purposes.

Riparian buffer lining a creek in Iowa.

Agricultural Wastewater Treatment

Non Point Source Controls

Sediment (loose soil) washed off fields is the largest source of agricultural pollution in the United States. Farmers may utilize erosion controls to reduce runoff flows and retain soil on their fields. Common techniques include contour plowing, crop mulching, crop rotation, planting perennial crops and installing riparian buffers.

Nutrients (nitrogen and phosphorus) are typically applied to farmland as commercial fertilizer, animal manure, or spraying of municipal or industrial wastewater (effluent) or sludge. Nutrients may also enter runoff from crop residues, irrigation water, wildlife, and atmospheric deposition. Farmers can develop and implement nutrient management plans to reduce excess application of nutrients and reduce the potential for nutrient pollution.

To minimize pesticide impacts, farmers may use Integrated Pest Management (IPM) techniques (which can include biological pest control) to maintain control over pests, reduce reliance on chemical pesticides, and protect water quality.

Feedlot in the United States

Point Source Wastewater Treatment

Farms with large livestock and poultry operations, such as factory farms, are called *concentrated animal feeding operations* or *feedlots* in the US and are being subject to increasing government regulation. Animal slurries are usually treated by containment in anaerobic lagoons before disposal by spray or trickle application to grassland. Constructed wetlands are sometimes used to facilitate treatment of animal wastes. Some animal slurries are treated by mixing with straw and composted at high temperature to produce a bacteriologically sterile and friable manure for soil improvement.

Erosion and Sediment Control from Construction Sites

Silt fence installed on a construction site.

Sediment from construction sites is managed by installation of:

- erosion controls, such as mulching and hydroseeding, and

- sediment controls, such as sediment basins and silt fences.

Discharge of toxic chemicals such as motor fuels and concrete washout is prevented by use of:

- spill prevention and control plans, and

- specially designed containers (e.g. for concrete washout) and structures such as overflow controls and diversion berms.

Control of Urban Runoff (Storm Water)

Effective control of urban runoff involves reducing the velocity and flow of storm water, as well as reducing pollutant discharges. Local governments use a variety of storm water management techniques to reduce the effects of urban runoff. These techniques, called best management practices (BMPs) in the U.S., may focus on water quantity control, while others focus on improving water quality, and some perform both functions.

Retention basin for controlling urban runoff

Pollution prevention practices include low-impact development techniques, installation of green roofs and improved chemical handling (e.g. management of motor fuels & oil, fertilizers and pesticides). Runoff mitigation systems include infiltration basins, bioretention systems, constructed wetlands, retention basins and similar devices.

Thermal pollution from runoff can be controlled by storm water management facilities that absorb the runoff or direct it into groundwater, such as bioretention systems and infiltration basins. Retention basins tend to be less effective at reducing temperature, as the water may be heated by the sun before being discharged to a receiving stream.

United States Regulation of Point Source Water Pollution

Point source water pollution comes from discrete conveyances and alters the chemical, biological, and physical characteristics of water. In the United States, it is largely regulated by the Clean Water Act (CWA). Among other things, the Act requires dischargers to obtain a National Pollutant Discharge Elimination System (NPDES) permit to legally discharge pollutants into a water body. However, point source pollution remains an issue in some water bodies, due to some limitations of the Act. Consequently, other regulatory approaches have emerged, such as water quality trading and voluntary community-level efforts.

Definition

Water pollution is the contamination of natural water bodies by chemical, physical, radioactive or pathogenic microbial substances. Point sources of water pollution are described by the CWA as "any discernible, confined, and discrete conveyance from which pollutants are or may be discharged." These include pipes or man-made ditches from stationary locations such as sewage treatment plants, factories, industrial wastewater treatment facilities, septic systems, ships, and other sources that are clearly discharging pollutants into water sources.

Relevant Science

The input of pollutants into a water body may impact the water's ability to deliver ecological, recreational, educational, and economic services. While the impacts of water pollution vary considerably based on a variety of site-specific factors, they may be either direct or indirect. Pollutants that are directly toxic pose a threat to organisms that may come into contact with contaminated water. These include persistent organic pollutants used as pesticides and toxic byproducts of industrial activity (such as cyanide). Other pollutants may indirectly impact ecosystem services by causing a change in water conditions that allows for a harmful activity to take place. This includes sediment (loose soil) inputs that decrease the amount of light that can penetrate through the water, reducing plant growth and diminishing oxygen availability for other aquatic organisms.

There are a variety of water quality parameters that may be affected by point source water pollution. They include: dissolved oxygen and biochemical oxygen demand (BOD), temperature, pH, turbidity, phosphorus, nitrates, total suspended solids, conductivity, alkalinity, and fecal coliform. Given that much of the point source water pollution in the United States comes from municipal wastewater treatment plants, BOD is perhaps the most widely used metric to assess water quality.

Water quality is also closely linked with water quantity issues. Water shortages from natural and anthropogenic activity reduce the dilutive properties of water and may concentrate water pollution. Oppositely, during flooding events, water pollution may spread to previously uncontaminated waters through surface overflow or the failure of man-made barriers.

Nature of the Problem/Context

Cuyahoga River fire

Before the CWA was enacted, companies indiscriminately discharged their effluents into water bodies. One such water body was the Cuyahoga River located in north-east Ohio. The river was thrust into the national limelight in 1969 when it caught fire, although the river had been plagued by fires since 1936. Pollution of the river had become prevalent in the early 1800s as contaminants from municipal and industrial discharges, bank erosion, commercial/residential development, atmospheric deposition, hazardous waste disposal sites, urban storm water runoff, combined sewer overflows (CSOs) and wastewater treatment plant bypasses were discharged into the river. *Time* magazine described the Cuyahoga as the river that "oozes rather than flows" and in which a person "does not drown but decays." The 1969 fire drew significant public attention to the state of the nation's waterways and is sometimes credited for the creation of the Federal Water Pollution Control Act (1972), commonly called the Clean Water Act, the Oil Pollution Act of 1990, and the establishment of federal and state agencies such as the Environmental Protection Agency (EPA).

Regulatory Framework

History of Regulation

Historically, regulation of point source water pollution in the United States included health- and use-based standards to protect environmental and economic interests. The Rivers and Harbors Act of 1899 contained provisions that made discharging refuse matter into navigable waters of the United States illegal without a permit issued by the U.S. Army Corps of Engineers.

In 1948 Congress passed the Federal Water Pollution Control Act (FWPCA). Although it was amended several times, the original FWPCA granted the Surgeon General of the Public Health Service the authority to develop programs to combat pollution that was harming surface and underground water sources. The FWPCA also authorized cooperation between federal and state agencies to construct waste treatment plants.

The FWPCA amendments of 1966, which came to be known as the *Clean Water Restoration Act* established a study to determine the effects of pollution on wildlife, recreation, and water supplies. The Act also set forth guidelines for abatement of water that may flow into international territory and prohibited the dumping of oil into navigable waters of the United States.

Further amendments to FWPCA in 1970 were dubbed the *Water Quality Improvement Act*. They required the development of certain water quality standards and expanded federal authority in upholding the standards. The most substantial amendments to the FWPCA occurred in 1972 and became known as the *Clean Water Act*.

Clean Water Act

Point source water pollution is largely regulated through the Clean Water Act, which gives the EPA the authority to set limits on the acceptable amount of pollutants that can be discharged into waters of the United States. The Act broadly defines a pollutant as any type of industrial, municipal, and agricultural waste discharged into water, such as: dredged soil, solid waste, incinerator residue, sewage, garbage, sewage sludge, munitions, chemical wastes, biological materials, radioactive materials, heat, wrecked or discarded equipment, rock, sand, cellar dirt and industrial, municipal, and agricultural waste. Point source water pollution is discharged into waters through both direct and indirect methods.

Indirect Point Source Water Pollution

An indirect discharger is one that sends its wastewater into a city sewer system, which carries it to the municipal sewage treatment plant or publicly owned treatment works (POTW). At the POTW, harmful pollutants in domestic sewage, called conventional pollutants, are removed from the sewage and then discharged into a surface water. POTWs are not designed to treat toxic or nonconventional pollutants in industrial wastewater, although they may incidentally remove some pollutants.

National Pretreatment Program

Indirect dischargers are covered by the National Pretreatment Program, which enforces three types of discharge standards:

- prohibited discharge standards – protect against pass-through and interference

- categorical standards – national, uniform, technology-based standards that limit the discharge of pollutants

- local limits – address the specific needs of a POTW and its receiving waters.

The goal of the pretreatment program is to protect municipal wastewater treatment plants from damage that may occur when hazardous, toxic, or other wastes are discharged into a sewer system and to protect the quality of sludge generated by these plants. Discharges to a POTW are regulated primarily by the POTW itself, rather than the state/tribe or EPA.

Direct Point Source Water Pollution

Deer Island Waste Water Treatment Plant, serving the Boston, Massachusetts area

Direct discharges are pollutants that are discharged directly into the water. To legally discharge pollutants directly into a waterbody, a National Pollution Discharge Elimination System (NPDES) permit must be obtained.

NPDES Permit Program

The NPDES permit program sets limits on the amount of pollutants that can be discharged into a waterbody. Technology based effluent limits establish a minimum level of pollution controls for all point source discharges. If technology based limits are not sufficient, water quality based effluent limits are developed.

Permits

Individual states are authorized by the EPA to issue permits when they have demonstrated that their program is at least as stringent as the EPA's program. States perform the day-to-day issuance of permits and oversight of the program while the EPA provides review and guidance to the states. All NPDES permits must contain a "specific, numeric, measurable set of limits on the amount of various pollutants that can appear in the wastewater discharged by the facility into the nation's waters" as well as guidelines on how often monitoring should be performed and what "sampling and analytic techniques should be used."

Types of Permits

- Individual – A unique permit is issued for each discharger.

- General – A single permit that covers a large number of similar dischargers in a specific geographic area. Examples include the EPA Vessels General Permit and industrial stormwater general permits.

Permitting Process

The authorized permit issuing body receives and reviews the permit application. The technology-based and water quality-based effluent limits are developed and then compared to determine which is the more stringent, which is then used as the effluent limit for the permit. Monitoring requirements, special conditions, and standard conditions for each pollutant are developed and the permit is then issued and its requirements are implemented.

- Technology-based requirements: A minimum level of treatment based on available treatment technologies is required for discharged pollutants, however, a discharger may use any available treatment technologies to meet the limits. The effluent limits are derived from different standards for different discharges:

  o Municipal discharges (POTW): National secondary treatment standards define limits of biological treatment standards based on biochemical oxygen demand (BOD), total suspended solids (TSS), and pH balance.

  o Non-municipal discharges: Limits for non-municipal (e.g. industrial) discharges are based on national discharge standards for industrial facilities within a certain category. These limits are achieved for using pollution control and prevention technologies designed for different types of dischargers. For existing dischargers, this level of treatment is equivalent to "Best Available Technology Economically Achievable" (BAT) and for new discharges, the treatment level is "New Source Performance Standards" (NSPS).

- Water quality-based requirements: Should technology-based standards not be stringent enough, Water Quality Based Effluent Limits (WQBEL) are developed to ensure that water quality standards are attained. WQBELs are based on ambient water quality standards.

Permit Components

All NPDES permits must contain the following five components:

- Cover page – indicates authorization for discharging and its locations

- Effluent limits – limits used to control discharges through technology-based or water quality-based standards

- Monitoring and reporting requirements – used to determine permit compliance

- Special conditions – can be used to supplement effluent limits

- Standard conditions – pre-established conditions that apply to all NPDES permits

Permit Violation

A permitee can be in violation of their permit when they discharge pollutants at a level higher than what is specified on their permit or discharge without a permit. They can also be in violation if they fail to comply with the monitoring and enforcement portion of the permit.

Enforcement

Since the NPDES permit program is a self-monitoring system where permitees are required to carry out detailed monitoring requirements, the EPA promotes "compliance assistance" as an enforcement technique, which "helps permittees come into, and remain, in compliance with their permit, rather than going immediately to enforcement actions." The EPA and state NPDES agencies have can perform periodic inspections and the EPA gives individual states the authority to enforce NPDES permits although the EPA has the right to carry out enforcement should a state not do so. Enforcement actions for violations include: injunctions, fines for typical violations, imprisonment for criminal violations, or supplemental environmental projects (SEP). Citizens may also bring suits against violators but they must first provide the EPA and state NPDES permit agency with the opportunity to take action.

Stormwater Management Permits

To address the nationwide problem of stormwater pollution, Congress broadened the CWA definition of "point source" in 1987 to include industrial stormwater discharges and municipal separate storm sewer systems ("MS4"). These facilities were required to obtain NPDES permits. This 1987 expansion was promulgated in two phases. The Phase I regulation, promulgated in 1990, required that all municipalities of 100,000 persons or more, industrial dischargers, and construction sites of 5 acres (20,000 m²) or more have NPDES permits for their stormwater discharges. Phase I permits were issued in much of the U.S. in 1991. The Phase II rule required that all municipalities, construction sites of 1 acre (4,000 m²) or more, and other large property owners (such as school districts) have NPDES permits for their stormwater discharges. EPA published the Phase II regulation in 1999.

A silt fence, a type of sediment control, installed on a construction site

Most construction sites are covered by general permits. Other industrial sites that only discharge stormwater are typically covered by general permits. Industrial stormwater dischargers that are otherwise required to have individual permits (due to their process wastewater and/or cooling water discharges), typically have the stormwater management requirements added to their individual permits.

In addition to implementing the NPDES requirements, many states and local governments have enacted their own stormwater management laws and ordinances, and some have published stormwater treatment design manuals. Some of these state and local requirements have expanded coverage beyond the federal requirements. For example, the State of Maryland requires erosion and sediment controls on construction sites of 5,000 sq ft (460 m²) or more. It is not uncommon for state agencies to revise their requirements and impose them upon counties and cities; daily fines ranging as high as $25,000 can be imposed for failure to modify their local stormwater permitting for construction sites, for instance.

Others

EPA Water Quality Trading Policy

The Clean Water Act has made great strides in reducing point source water pollution, but this effect is overshadowed by the fact that nonpoint source pollution, which is not subject to regulation under the Act, has correspondingly increased. One of the solutions to address this imbalance is point/nonpoint source trading of pollutants. In January 2003, the EPA Water Quality Trading Policy was issued. At this time, many waters in the United States did not support their designated uses. Specifically, 40 percent of rivers, 45 percent of streams, and 50 percent of lakes that had been surveyed were unfit. Consequently, when The Water Quality Trading Policy was issued it acknowledged that "the progress made toward restoring and maintaining the chemical, physical, and biological integrity of the nation's waters under the 1972 Clean Water Act and its National Pollutant Discharge Elimination System (NPDES) permits has been incomplete."

The purpose of the policy is to "encourage voluntary trading programs that facilitate the implementation of TMDLs, reduce the costs of compliance with CWA regulations, establish incentives for voluntary reductions, and promote watershed-based initiatives (3)." The policy supports the trading of nutrients such as nitrogen and phosphorus and sediment load reductions, but in order for it to be extended to other contaminants, more scrutiny is required. All water quality trading programs are subject to the requirements of the Clean Water Act.

The Trading Policy outlines basic ground rules for trading by specifying viable pollutants, how to set baselines, and detailing the components of credible trading programs. It also stipulates that trades must occur within the same watershed. Water quality trading programs are subject to the stipulations of the Clean Water Act.

Other Laws that May Affect Some NPDES Permits

- Endangered Species Act: Federal agencies must consult with the U.S. Fish and Wildlife Service to ensure that discharges from a project (e.g., a new or expanded industrial facility) will not endanger a threatened species or their habitat.

- National Environmental Policy Act (NEPA): Only discharges that are subject to New Source Performance Standards (or new sources otherwise defined in the NPDES regulations) are subject to NEPA review prior to being issued a permit.

- National Historic Preservation Act: Prior to issuing a permit, EPA Regional Administrators must adopt measures that mitigate adverse effects on properties in the National Register of Historic Places.

- Coastal Zone Management Act: Permits will not be issued unless the permitees certifies that proposed activities, which would affect land or water use in coastal zones, comply with the Coastal Zone Management Act.

- Wild and Scenic Rivers Act: Prohibits issuance of permit for water resources projects that will have a direct, adverse effect on the values for which a national wild and scenic river was established.

- Fish and Wildlife Coordination Act: Jurisdiction over wildlife resources must be established prior to permit issuance so that resources can be conserved.

- *Magnuson-Stevens Act:* The "Essential Fish Habitat Provisions" in the law require EPA to consult with the National Marine Fisheries Service for any EPA-issued permits which may adversely affect essential fish habitat.

Problems/Issues/Concerns

Funding

Cost issues for monitoring

Monitoring of water bodies is the responsibility of authorized states, not the EPA. In 1997, EPA estimated that private and public point source control costs were $14 billion and $34 billion, respectively. The EPA has acknowledged that states have not adequately funded their monitoring programs, which has led to some uncertainty regarding the quality of most surface waters.

Enforcement

Self-monitoring and self-reporting

In many cases, the enforcement mechanisms of the Clean Water Act have created tension between regulators, regulated parties, and local citizens. Most NPDES permits require dischargers to monitor and report the contents of their discharges to the appropriate authorities. This requirement is potentially self-incriminating, forcing industries to provide information that may subject them to penalties and legal constraints. As a result, some dischargers go to great lengths to avoid penalties, including falsifying discharge monitoring reports and tampering with monitoring equipment. In *United States v. Hopkins*, the court ruled on a case where the vice president for manufacturing at Spirol International Corporation was charged with three criminal violations for falsifying water samples sent to state regulatory agencies. Spirol diluted his samples, which contained high levels of zinc, with tap water on numerous occasions and frequently ordered his subordinates to reduce the zinc concentration in the water by running it through a coffee filter.

Tensions between state and federal government

Like other environmental laws, the Clean Water Act delegates many enforcement responsibilities to state agencies. While the burden of enforcement may be transferred to the states, federal agencies reserve the right to approve or reject state plans for dealing with water pollution. This relationship reduces the regulatory burden on federal agencies, but can lead to confusion and tension between the two regulators.

Many of these tensions arise with regards to the commerce clause of the constitution. Until recently, the commerce clause has given the federal government considerable authority in regulating states' decisions about water use. In 2000, the United States Supreme Court ruled on *Solid Waste Agency of Northern Cook County v. US Army Corps of Engineers*. This ruling struck down the Corps' ability to prevent the construction of a disposal site for non-hazardous waste in Illinois based on power derived from the commerce clause. The Corps cited the Migratory Bird Rule when they initially denied the section 404 permit under the Clean Water Act. The migratory bird rule was meant to protect habitats used by migratory birds, which included the abandoned mining site that SWANCC had proposed to construct the waste disposal site. Chief Justice Rehnquist wrote: "Congress passes the CWA for the state purpose of 'restoring and maintaining the chemical, physical, and biological integrity of the Nation's waters.' In doing so,, Congress chose to recognize, preserve, and protect the primary responsibilities and rights of States to prevent, reduce, and eliminate pollution, to plan the development and use... of land and water resources...". In reversing the Corps' decision to issue a permit, the court reversed a trend and placed a check on federal power over state land use and water rights. Tensions between federal and state agencies concerning interstate commerce and point source water pollution continue, and are a reality of the Clean Water Act.

Ambiguity of the CWA

Coeur Alaska v. Southeast Alaska Conservation Council

In 2009, the Supreme Court ruled on *Coeur Alaska, Inc. v. Southeast Alaska Conservation Council*. The case concerned the re-opening of a gold mine outside Juneau, Alaska that had been out of operation since 1928. Coeur Alaska planned to utilize froth flotation in order to extract gold, creating 4.5 million tons of tailings over the course of its lifetime. The mining company opted to dispose of the tailings in nearby Lower Slate Lake, requiring a permit to comply with the Clean Water Act. The tailings would fundamentally change the physical and chemical characteristics of the lake, raising the lake bed by 50 feet and expanding the area from 23 to 60 acres. Coeur Alaska proposed to temporarily re-route nearby streams around Lower Slate Lake until they could purify the water and re-introduce the natural flow patterns.

Tailings from froth flotation contain high concentrations of heavy metals, including aluminum, which have toxic effects of aquatic organisms. As a result, the disposal of these tailings into Lower Slate Lake is eligible for a section 402 permit for discharge of a pollutant from the EPA (NPDES permit). The nature of the tailings also justifies their categorization as a fill material, or a "material [that] has the effect of... changing the bottom elevation" of a water body. Consequently, Coeur Alaska was also eligible for a "Dredge and Fill" permit from the Army Corps of Engineers under CWA section 404. The company applied for this latter permit and received authorization from the Corps to dump the tailings into Lower Slate Lake. The Southeast Alaska Conservation Council contended that disposal of the tailings is explicitly banned by section 306(e) of the Clean Water Act, and would therefore make Coeur Alaska ineligible for a NPDES permit.

The Court ruled in favor of Coeur Alaska, explaining that if the Army Corps of Engineers has authority to issue a permit under section 404, the EPA does not have authority to issue a section 402 permit. They asserted that the law is ambiguous as to whether section 306 applies to fill materials and found no erroneous or unreasonable behavior by the Corps. As a result, although the tailings would explicitly violate the Clean Water Act under section 402, the Corps may issue a dredge and fill permit.

This decision has not resonated well with environmental groups, who are worried that the decision may allow companies to discharge massive amounts of hazardous pollutants by avoiding the NPDES permitting procedure. Of particular concern is the mountaintop mining industry, which has the capacity to fundamentally alter aquatic ecosystems by filling in water bodies with sediment and mining debris. This tension between various sections within the Clean Water Act is sure to receive considerable attention in future years.

Other Emerging Regulatory Approaches

Water Quality Trading

Water quality trading (WQT) is a market-based approach, implemented on a watershed-scale, used to improve or maintain water quality. It involves the voluntary exchange of pollution reduction credits from sources with low costs of pollution control to those with high costs of pollution control. WQT programs are still subject to the requirements of the Clean Water Act, but they can be used to reduce the overall cost of compliance. Usually, permitted point sources of water pollution, such as wastewater treatment plants, have high discharge treatment costs, whereas nonpoint sources of water pollution, such as agriculture, have low costs of pollution reduction. Therefore, it is generally assumed that most trades would take place between point sources and nonpoint sources. However, point source-point source trades could also occur as well as pretreatment trades and intra-plant trades.

Most of the water quality trading markets currently in operation are focused on the trading of nutrients such as phosphorus and nitrogen. However, increasing interest has been shown in trading programs for sediment runoff, biological oxygen demand, and temperature. WQT programs can be used to preserve good water quality in unimpaired waters by counterbalancing new or increased pollutant discharges. In impaired waters, a WQT program can be used to improve water quality by reducing pollutant discharges in order to meet a specified water quality standard or total maximum daily load (TMDL).

TMDLs apply to both point sources and nonpoint sources and they represent the primary impetus for WQT programs. Point sources of pollutants that require NPDES permits often have strict discharge limits based on a TMDL. WQT can allow these sources to obtain lower costs of compliance, while still achieving the overall desired pollution reduction. Several factors influence whether or not a TMDL-based water quality trading program will be successful. First, the market must be appropriately structured within the regulatory framework of the Clean Water Act. Second, the pollutant must be well-suited for trading. Third, implementation of a WQT market requires public input and voluntary participation. Finally, there must be adequate differences in pollution control costs and available opportunities for reduction.

Credits and trade ratios

In a WQT market, a unit of pollutant reduction is called a credit. A point source can generate credits by reducing its discharge below its most stringent effluent limitation and a nonpoint source can generate credits "by installing best management practices (BMPs) beyond its baseline". Before being able to purchase credits, source must first meet its technology-based effluent limit (TBEL). The credits can then be used to meet water quality-based effluent limits (WQBEL).

In order to ensure that trades are effective and do not result in more pollution than would occur in their absence, trade ratios are used. Trade ratios can have several components including:

- Location: Source location relevant to the downstream area of concern can be an important factor.

- Delivery: The distance between sources can play a role in determining whether permit requirements are met at the outfall.

- Uncertainty: Nonpoint source reductions can be difficult to quantify.

- Equivalency: Sources may be discharging different forms of the same pollutant.

- Retirement: Credits may be retired to achieve further water quality improvement.

Permitted point sources can trade with other point sources or nonpoint sources. Trades can occur directly, or be brokered by third parties. However, when dealing with nonpoint source reductions, a level of uncertainty does exist. In order to address this, monitoring should be conducted. Modeling can also be used as a supplement to monitoring. Uncertainty can also be mitigated by field testing BMPs and using conservative assumptions for BMP efficacy.

Benefits

There are many economic, environmental, and social benefits that can be gained by establishing a WQT market within a watershed. Economically, since WQT is a market-based policy instrument, substantial savings can be generated while still achieving a mandated water quality goal. According to The National Cost to Implement Total Maximum Daily Loads (TMDLs) Draft report, flexible approaches to improving water quality, such as WQT markets, could save $900 million per year when compared with the least flexible approach (3). In 2008, WQT programs were worth $11 million, but have the potential for rapid growth. Other economic benefits of WQT include a reduction in the overall costs of compliance, the ability for dischargers to take advantages of economies of scale and differences in treatment efficiencies, and the ability to maintain growth without further harming the environment.

The environmental benefits of WQT programs are also numerous. First, habitats and ecosystems are protected and/or improved. Second, water quality objectives are able to be achieved in a timely manner. Third, there is incentive for innovation and creation of pollution prevention technologies. Finally, nonpoint sources are included in solving water quality problems. Social benefits include dialog among watershed stakeholders and incentives for all dischargers to reduce their pollution.

Factors influencing success

The success of a WQT market is determined by several factors including the pollutant of interest, physical characteristics of the affected watershed, control costs, trading mechanisms, and stakeholder participation and willingness. It is also important that the desired level of pollution reduction is not so great that all sources must reduce the maximum amount possible because this would eliminate surplus reductions to be used for credits.

Obstacles to implementation

The biggest obstacle to the widespread adoption of WQT markets is lack of supply and demand. In fact, studies of current water quality trading programs indicate that the typical problems associated with inhibition of water quality trading, such as high transaction costs, poor institutional infrastructure, and uncertain criteria, are being overcome. The main problem is that, under existing regulatory conditions, there are simply not enough willing buyers and sellers. Currently, most nonpoint sources of water pollution are unregulated or, assuming detection occurs, have relatively small consequences for violations. Consequently, nonpoint sources do not have incentive to participate in WQT. For WQT markets to be successful, greater demand is needed for pollution credits. For this to happen, water quality standards need to be clear and enforceable.

Contrast to emissions trading

Given the success of the sulfur dioxide emission trading market that was established to combat acid rain, at first glance it seems that this level of success should be easily extended to water quality trading. However, the reason this has not occurred yet comes down to a fundamental difference between water pollution and air pollution, and the process of establishing their respective markets. Establishing an emissions market, in principle, has three steps: (1) set a cap on emissions, (2) allocate portions of the cap to individual firms, and (3) allow each firm to meet its allowance through emission reduction or trade. The difference with water pollution, however, is that the problems that cause local water quality issues differ from those that create regional air pollution problems. Discharges into water are difficult to measure and effects are dependent on a variety of other factors and vary with weather and location.

Dealing with Transboundary Pollution: A Case Study of the U.S./ Mexico Border Region

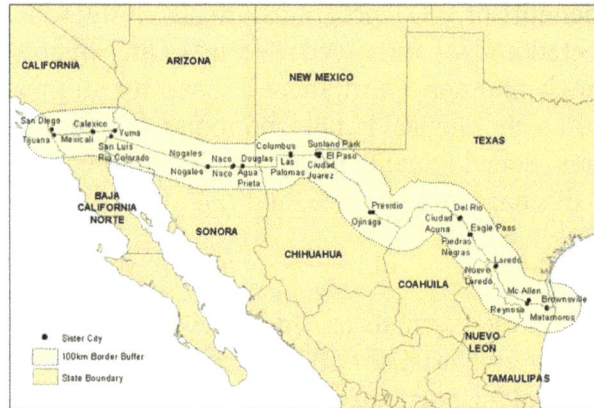

Map of the U.S. – Mexico border region

Overview

The border region (approximately 2,000 miles (3,200 km) long and 62 miles (100 km) wide) is predominantly arid and contains seven watersheds including the Rio Grande which forms part of the border. The watersheds provide numerous benefits for the 14 cities and over nine million people in the region. However, increasing population, the arid climatic conditions of the region, the nature of economic activities along the rivers, increased trade, and uncontrolled emissions into them have placed tremendous pressure on water resources and threatened natural ecosystems. A large proportion of the population lacks access to clean drinking water and sanitation triggering public health concerns.

Policy Issues

Point source water pollution is a source of concern along the US-Mexico border as pollutants from both countries are entering shared waterways due to agricultural runoff, industrial discharge, and untreated sewage. Various policy issues arise in attempting to deal with this and include:

- Some pollution originates from areas beyond the border region as pollution is carried into the region by the waterways. This makes it difficult to regulate as discharges are difficult to apportion and control.

- Pollution is caused by and affects both countries therefore requiring a joint response.

- The socio-economic differences between the two countries have implications for policy implementation and enforcement.

- Various interests are represented with strong influence from environmental and social groups. Multiple levels of government agencies are also involved.

Policy Responses

There have been several attempts to address environmental concerns in the border region in the past by both governments. Significant intervention, however, resulted from the North American Free Trade Agreement (NAFTA) of 1994 between the U.S., Canada, and Mexico which renewed concerns over the environmental quality of the region due to increased trade in the region. The two governments therefore entered into the US-Mexico Border Environment Cooperation Agreement which created a number of institutions and programs. The Border Environmental Cooperation Commission(-BECC) and the North American Development Bank (NADB) were created to address border environmental-infrastructure issues and were effected in 2004. Distinct characteristics of these institutions and their approach are that they: are truly bi-national (have members from both countries) at all levels; emphasize a bottom-up approach with enhanced public participation; have a preference to assist disadvantaged communities; avoid regulatory or standard-driven approaches; emphasize economic and environmental sustainability.

The US-Mexico Border Program was also created by the agreement and placed under the management of the EPA (under regions 6 and 9) to correct the oversights of previous institutions and give guidance to cross-border environmental policy. The three institutions work together to identify, develop, finance and implement projects in the communities and certify them as "environmentally sustainable" subsequently funding them through a grant-making process. Communities, public, and private entities (sponsors) are invited to submit water and wastewater infrastructure projects. These projects are required to meet certain criteria to qualify for certification and funding. Among other requirements, they have to address an eligible environmental sector; must have a U.S.-side benefit; and have adequate planning, operations and maintenance, and pretreatment provisions. One specific provision touching on point-source water pollution states that "projects where the discharge is directly or indirectly to U.S. side waters, must target achievement of U.S. norms for ambient water quality in U.S. side waters, although infrastructure development may be phased over time. Any flow reductions that result from implementation of non-discharging alternatives must not threaten U.S. or shared ecosystems". Projects receive significant input from the communities living in the region in determining their sustainability. After certification, the project then receives funding from the NADB. So far, the BECC has certified 160 projects worth approximately three billion dollars since 2004.

The border program has also facilitated direct provision of infrastructure by the federal and state governments such as the construction of wastewater treatment plants, sewer lines, and raw water storage lagoons. One such example is the construction of the Matamoros lift station which is the first phase in eliminating raw sewage discharges into the Rio Grande. Further, the program emphasizes the provision of environmental education and information to communities living in the region. The EPA has established an informational website to provide news and information on the program.

Non-point Source Pollution

Muddy river

Nonpoint source (NPS) pollution is a term used to describe pollution resulting from many diffuse sources, in direct contrast to point source pollution which results from a single source. Nonpoint source pollution generally results from land runoff, precipitation, atmospheric deposition, drainage, seepage, or hydrological modification (rainfall or snowmelt) where tracing the pollution back to a single source is difficult.

Non-point source water pollution affects a water body from sources such as polluted runoff from agricultural areas draining into a river, or wind-borne debris blowing out to sea. Non-point source air pollution affects air quality from sources such as smokestacks or car tailpipes. Although these pollutants have originated from a point source, the long-range transport ability and multiple sources of the pollutant make it a non-point source of pollution. Non-point source pollution can be contrasted with point source pollution, where discharges occur to a body of water or into the atmosphere at a single location.

NPS may derive from many different sources with no specific solution may change to rectify the problem, making it difficult to regulate. Non point source water pollution is difficult to control because it comes from the everyday activities of many different people, such as fertilizing a lawn, using a pesticide, or constructing a road or building.

It is the leading cause of water pollution in the United States today, with polluted runoff from agriculture the primary cause.

Other significant sources of runoff include hydrological and habitat modification, and silviculture (forestry).

Contaminated stormwater washed off parking lots, roads and highways, and lawns (often containing fertilizers and pesticides) is called urban runoff. This runoff is often classified as a type of NPS pollution. Some people may also consider it a point source because many times it is channeled into municipal storm drain systems and discharged through pipes to nearby surface waters. However, not all urban runoff flows through storm drain systems before entering water bodies. Some may flow directly into water bodies, especially in developing and suburban areas. Also, unlike other types of point sources, such as industrial discharges, sewage treatment plants and other operations, pollution in urban runoff cannot be attributed to one activity or even group of activities. Therefore, because it is not caused by an easily identified and regulated activity, urban runoff pollution sources are also often treated as true non-point sources as municipalities work to abate them.

Principal Types

Runoff of soil and fertilizer during a rain storm

Sediment

Sediment (loose soil) includes silt (fine particles) and suspended solids (larger particles). Sediment may enter surface waters from eroding stream banks, and from surface runoff due to improper plant cover on urban and rural land. Sediment creates turbidity (cloudiness) in water bodies, reducing the amount of light reaching lower depths, which can inhibit growth of submerged aquatic plants and consequently affect species which are dependent on them, such as fish and shellfish. High turbidity levels also inhibit drinking water purification systems.

Sediment can also be discharged from multiple different sources. Sources include construction sites (although these are point sources, which can be managed with erosion controls and sediment controls), agricultural fields, stream banks, and highly disturbed areas.

Nutrients

Nutrients mainly refers to inorganic matter from runoff, landfills, livestock operations and crop lands. The two primary nutrients of concern are phosphorus and nitrogen.

Phosphorus is a nutrient that occurs in many forms that are bioavailable. It is notoriously over-abundant in human sewage sludge. It is a main ingredient in many fertilizers used for agriculture as well as on residential and commercial properties, and may become a limiting nutrient in freshwater systems and some estuaries. Phosphorus is most often transported to water bodies via soil erosion because many forms of phosphorus tend to be adsorbed on to soil particles. Excess amounts of phosphorus in aquatic systems (particularly freshwater lakes, reservoirs, and ponds) leads to proliferation of microscopic algae called phytoplankton. The increase of organic matter supply due to the excessive growth of the phytoplankton is called eutrophication. A common symptom of eutrophication is algae blooms that can produce unsightly surface scums, shade out beneficial types of plants, and poison the water due to toxins produced by the algae. These toxins are a particular problem in systems used for drinking water because some toxins can cause human illness and removal of the toxins is difficult and expensive. Bacterial decomposition of algal blooms consumes dissolved oxygen in the water, generating hypoxia with detrimental consequences for fish and aquatic invertebrates.

Nitrogen is the other key ingredient in fertilizers, and it generally becomes a pollutant in saltwater or brackish estuarine systems where nitrogen is a limiting nutrient. Similar to phosphorus in fresh-waters, excess amounts of bioavailable nitrogen in marine systems lead to eutrophication and algae blooms. Hypoxia is an increasingly common result of eutrophication in marine systems and can impact large areas of estuaries, bays, and near shore coastal waters. Each summer, hypoxic conditions form in bottom waters where the Mississippi River enters the Gulf of Mexico. During recent summers, the aereal extent of this "dead zone" is comparable to the area of New Jersey and has major detrimental consequences for fisheries in the region.

Nitrogen is most often transported by water as nitrate (NO_3). The nitrogen is usually added to a watershed as organic-N or ammonia (NH_3), so nitrogen stays attached to the soil until oxidation converts it into nitrate. Since the nitrate is generally already incorporated into the soil, the water traveling through the soil (i.e., interflow and tile drainage) is the most likely to transport it, rather than surface runoff.

Toxic Contaminants and Chemicals

Compounds including heavy metals like lead, mercury, zinc, and cadmium, organics like polychlorinated biphenyls (PCBs) and polycyclic aromatic hydrocarbons (PAHs), fire retardants, and other substances are resistant to breakdown. These contaminants can come from a variety of sources including human sewage sludge, mining operations, vehicle emissions, fossil fuel combustion, urban runoff, industrial operations and landfills.

Toxic chemicals mainly include organic compounds and inorganic compounds. These compounds include pesticides like DDT, acids, and salts that have severe effects to the ecosystem and water-bodies. These compounds can threaten the health of both humans and aquatic species while being resistant to environmental breakdown, thus allowing them to persist in the environment. These toxic chemicals could come from croplands, nurseries, orchards, building sites, gardens, lawns and landfills.

Acids and salts mainly are inorganic pollutants from irrigated lands, mining operations, urban runoff, industrial sites and landfills.

Pathogens

Pathogens are bacteria and viruses that can be found in water and cause diseases in humans. Typically, pathogens cause disease when they are present in public drinking water supplies. Pathogens found in contaminated runoff may include:

- *Cryptosporidium parvum*

- *Giardia lamblia*

- *Salmonella*

- *Novovirus* and other viruses

- Parasitic worms (helminths).

Coliform bacteria and fecal matter may also be detected in runoff. These bacteria are a commonly used indicator of water pollution, but not an actual cause of disease.

Pathogens may contaminate runoff due to poorly managed livestock operations, faulty septic systems, improper handling of pet waste, the over application of human sewage sludge, contaminated storm sewers, and sanitary sewer overflows.

Principal Sources

Urban and Suburban Areas

Urban and suburban areas are a main sources of nonpoint source pollution due to the amount of runoff that is produced due to the large amount of paved surfaces. Paved surfaces, such as asphalt and concrete are impervious to water penetrating them. Any water that is on contact with these surfaces will run off and be absorbed by the surrounding environment. These surfaces make it easier for stormwater to carry pollutants into the surrounding soil.

Construction sites tend to have disturbed soil that is easily eroded by precipitation like rain, snow, and hail. Additionally, discarded debris on the site can be carried away by runoff waters and enter the aquatic environment.

Typically, in suburban areas, chemicals are used for lawn care. These chemicals can end up in runoff and enter the surrounding environment via storm drains in the city.

Since the water in storm drains is not treated before flowing into surrounding water bodies, the chemicals enter the water directly.

Agricultural Operations

Agricultural operations account for a large percentage of all nonpoint source pollution in the United States. When large tracts of land are plowed to grow crops, it exposes and loosens soil that was once buried. This makes the exposed soil more vulnerable to erosion during rainstorms. It also can increase the amount of fertilizer and pesticides carried into nearby bodies of water.

Atmospheric Inputs

Atmospheric inputs of pollution into the air can come from multiple sources. Typically, industrial facilities, like factories, emit air pollution via a smokestack. Although this is a point source, due to the distributional nature, long-range transport, and multiple sources of the pollution, it is considered a nonpoint source. Additionally, atmospheric pollution can become water pollution, by being washed out of the atmosphere in the form of rain or snow.

Highway Runoff

Highway runoff accounts for a small but widespread percentage of all nonpoint source pollution. Harned (1988) estimated that runoff loads were composed of atmospheric fallout (9%), vehicle deposition (25%) and highway maintenance materials (67%) he also estimated that about 9 percent of these loads were reentrained in the atmosphere.

Forestry and Mining Operations

Forestry and mining operations can have significant inputs to non-point source pollution.

Forestry

Forestry operations reduce the number of trees in a given area, thus reducing the oxygen levels in that area as well. This action, coupled with the heavy machinery rolling over the soil increases the risk of erosion.

Mining

Active mining operations are considered point sources, however runoff from abandoned mining operations contribute to nonpoint source pollution. In strip mining operations, the top of the mountain is removed to expose the desired ore. If this area is not properly reclaimed once the mining has finished, soil erosion can occur. Additionally, there can be chemical reactions with the air and newly exposed rock to create acidic runoff.

Water that seeps out of abandoned subsurface mines can also be highly acidic. This can seep into the nearest body of water and change the pH in the aquatic environment.

Marinas and Boating Activities

Chemicals used for boat maintenance, like paint, solvents, and oils find their way into water through runoff. Additionally, spilling fuels or leaking fuels directly into the water from boats contribute to nonpoint source pollution. Nutrient and bacteria levels are increased by poorly maintained sanitary waste receptacles on the boat and pump-out stations.

Control

Contour buffer strips used to retain soil and reduce erosion

Urban and Suburban Areas

To control non-point source pollution, many different approaches can be undertaken in both urban and suburban areas. Buffer strips provide a barrier of grass in between impervious paving material like parking lots and roads, and the closest body of water. This allows the soil to absorb any pollution before it enters the local aquatic system. Retention ponds can be built in drainage areas to create an aquatic buffer between runoff pollution and the aquatic environment. Runoff and storm water drain into the retention pond allowing for the contaminants to settle out and become trapped in the pond. The use of porous pavement allows for rain and storm water to drain into the ground beneath the pavement, reducing the amount of runoff that drains directly into the water body. Restoration methods such as constructing wetlands are also used to slow runoff as well as absorb contamination.

Construction sites typically implement simple measures to reduce pollution and runoff. Firstly, sediment or silt fences are erected around construction sites to reduce the amount of sediment and large material draining into the nearby water body. Secondly, laying grass or straw along the border of construction sites also work to reduce nonpoint source pollution.

In areas served by single-home septic systems, local government regulations can force septic system maintenance to ensure compliance with water quality standards. In

Washington (state), a novel approach was developed through a creation of a "shellfish protection district" when either a commercial or recreational shellfish bed is downgraded because of ongoing nonpoint source pollution. The shellfish protection district is a geographic area designated by a county to protect water quality and tideland resources, and provides a mechanism to generate local funds for water quality services to control nonpoint sources of pollution. At least two shellfish protection districts in south Puget Sound have instituted septic system operation and maintenance requirements with program fees tied directly to property taxes.

Agricultural operations

To control sediment and runoff, farmers may utilize erosion controls to reduce runoff flows and retain soil on their fields. Common techniques include contour plowing, crop mulching, crop rotation, planting perennial crops and installing riparian buffers. *Conservation tillage* is a concept used to reduce runoff while planting a new crop. The farmer leaves some crop reside from the previous planting in the ground to help prevent runoff during the planting process.

Nutrients are typically applied to farmland as commercial fertilizer; animal manure; or spraying of municipal or industrial wastewater (effluent) or sludge. Nutrients may also enter runoff from crop residues, irrigation water, wildlife, and atmospheric deposition. Farmers can develop and implement nutrient management plans to reduce excess application of nutrients.

To minimize pesticide impacts, farmers may use Integrated Pest Management (IPM) techniques (which can include biological pest control) to maintain control over pests, reduce reliance on chemical pesticides, and protect water quality.

Forestry Operations

With a well-planned placement of both logging trails, also called skid trails, can reduce the amount of sediment generated. By planning the trails location as far away from the logging activity as possible as well as contouring the trails with the land, it can reduce the amount of loose sediment in the runoff. Additionally, by replanting trees on the land after logging, it provides a structure for the soil to regain stability as well as replaces the logged environment.

Marinas

Installing shut off valves on fuel pumps at a marina dock can help reduce the amount of spillover into the water. Additionally, pump-out stations that are easily accessible to boaters in a marina can provide a clean place in which to dispose of sanitary waste without dumping it directly into the water. Finally, something as simple as having trash containers around a marina can prevent larger objects entering the water.

Groundwater Pollution

Groundwater pollution example in Lusaka, Zambia where the pit latrine in the background is polluting the shallow well in the foreground with pathogens and nitrate.

Groundwater pollution (also called groundwater contamination) occurs when pollutants are released to the ground and make their way down into groundwater. It can also occur naturally due to the presence of a minor and unwanted constituent, contaminant or impurity in the groundwater, in which case it is more likely referred to as contamination rather than pollution.

The pollutant creates a contaminant plume within an aquifer. Movement of water and dispersion within the aquifer spreads the pollutant over a wider area. Its advancing boundary, often called a plume edge, can intersect with groundwater wells or daylight into surface water such as seeps and spring, making the water supplies unsafe for humans and wildlife. The movement of the plume, called a plume front, may be analyzed through a hydrological transport model or groundwater model. Analysis of groundwater pollution may focus on soil characteristics and site geology, hydrogeology, hydrology, and the nature of the contaminants.

Pollution can occur from on-site sanitation systems, landfills, effluent from wastewater treatment plants, leaking sewers, petrol filling stations or from over application of fertilizers in agriculture. Pollution (or contamination) can also occur from naturally occurring contaminants, such as arsenic or fluoride. Using polluted groundwater causes hazards to public health through poisoning or the spread of disease.

Different mechanisms have influence on the transport of pollutants, e.g. diffusion, adsorption, precipitation, decay, in the groundwater. The interaction of groundwater contamination with surface waters is analyzed by use of hydrology transport models.

Pollutant Types

Contaminants found in groundwater cover a broad range of physical, inorganic chemical, organic chemical, bacteriological, and radioactive parameters. Principally, many of the same pollutants that play a role in surface water pollution may also be found in polluted groundwater, although their respective importance may differ.

Arsenic and Fluoride

Arsenic and fluoride have been recognized by the World Health Organization (WHO) as the most serious inorganic contaminants in drinking-water on a worldwide basis.

The metalloid arsenic can occur naturally in groundwater, as seen most frequently in Asia, including in China, India and Bangladesh. In the Ganges Plain of northern India and Bangladesh severe contamination of groundwater by naturally occurring arsenic affects 25% of water wells in the shallower of two regional aquifers.

Arsenic in groundwater can also be present where there are mining operations or mine waste dumps that will leach arsenic.

Natural fluoride in groundwater is of growing concern as deeper groundwater is being used, "with more than 200 million people at risk of drinking water with elevated concentrations." Fluoride can especially be released from acidic volcanic rocks and dispersed volcanic ash when water hardness is low. High levels of fluoride in groundwater is a serious problem in the Argentinean Pampas, Chile, Mexico, India, Pakistan, the East African Rift, and some volcanic islands (Tenerife)

In areas that have naturally occurring high levels of fluoride in groundwater which is used for drinking water, both dental and skeletal fluorosis can be prevalent and severe.

Pathogens

Waterborne diseases can be spread via a groundwater well which is contaminated with fecal pathogens from pit latrines

Pathogens contained in feces can lead to groundwater pollution when they are given the opportunity to reach the groundwater, making it unsafe for drinking. Of the four pathogen types that are present in feces (bacteria, viruses, protozoa and helminths or helminth eggs), the first three can be commonly found in polluted groundwater, whereas the relatively large helminth eggs are usually filtered out by the soil matrix.

Groundwater that is contaminated with pathogens can lead to fatal fecal-oral transmission of diseases (e.g. cholera, diarrhoea).

Nitrate

Nitrate is the most common chemical contaminant in the world's groundwater and aquifers. In some low-income countries nitrate levels in groundwater are extremely high, causing significant health problems. It is also stable (it does not degrade) under high oxygen conditions.

Nitrate levels above 10 mg/L (10 ppm) in groundwater can cause "blue baby syndrome" (acquired methemoglobinemia). Drinking water quality standards in the European Union stipulate less than 50 mg/L for nitrate in drinking water.

However, the linkages between nitrates in drinking water and blue baby syndrome have been disputed in other studies. The syndrome outbreaks might be due to other factors than elevated nitrate concentrations in drinking water.

Elevated nitrate levels in groundwater can be caused by on-site sanitation, sewage sludge disposal and agricultural activities. It can therefore have an urban or agricultural origin.

Organic Compounds

Volatile organic compounds (VOCs) are a dangerous contaminant of groundwater. They are generally introduced to the environment through careless industrial practices. Many of these compounds were not known to be harmful until the late 1960s and it was some time before regular testing of groundwater identified these substances in drinking water sources.

Primary VOC pollutants found in groundwater include aromatic hydrocarbons such as BTEX compounds (benzene, toluene, ethylbenzene and xylenes), and chlorinated solvents including tetrachloroethylene (PCE), trichloroethylene (TCE), and vinyl chloride (VC). BTEX are important components of gasoline. PCE and TCE are industrial solvents historically used in dry cleaning processes and as a metal degreaser, respectively.

Other organic pollutants present in groundwater and derived from industrial operations are the polycyclic aromatic hydrocarbons (PAHs). Due to its molecular weight, Naphthalene is the most soluble and mobile PAH found in groundwater, whereas benzo(a)pyrene is the most toxic one. PAHs are generally produced as byproducts by in-complete combustion of organic matter.

Organic pollutants can also be found in groundwater as insecticides and herbicides. As many other synthetic organic compounds, most pesticides have very complex molec-ular structures. This complexity determines the water solubility, adsorption capacity, and mobility of pesticides in the groundwater system. Thus, some types of pesticides are more mobile than others so they can more easily reach a drinking-water source.

Metals

Several trace metals can occurs naturally in some rocks or enter in the environmen-tal from natural processes such as weathering. However, industrial activities such as mining and metallurgy, solid waste disposal, painting and enamel works can lead to severe groundwater pollution with elevated concentrations of toxic metals including lead, cadmium and chromium.

The migration of metals (and metalloids) in groundwater will be affected by several factors, in particular by chemical reactions which determine the partitioning of con-taminants among different phases and species. Thus, the mobility of metals primarily depends on the pH and redox state of groundwater.

Others

Other organic pollutants include a range of organohalides and other chemical com-pounds, petroleum hydrocarbons, various chemical compounds found in personal hy-giene and cosmetic products, drug pollution involving pharmaceutical drugs and their metabolites. Inorganic pollutants might include other nutrients such as ammonia and phosphate, and radionuclides such as uranium (U) or radon (Rn) naturally present in some geological formations. Saltwater intrusion is also an example of natural contam-ination, but is very often intensified by human activities.

Groundwater pollution is a worldwide issue. A study of the groundwater quality of the principal aquifers of the United States conducted between 1991 and 2004, showed that 23% of domestic wells had contaminants at levels greater than human-health bench-marks. Another study suggested that the major groundwater pollution problems in Africa, considering the order of importance are: (1) nitrate pollution, (2) pathogenic agents, (3) organic pollution, (4) salinization, and (5) acid mine drainage.

Causes

Naturally-occurring (Geogenic)

"Geogenic" refers to naturally occurring as a result from geological processes.

The natural arsenic pollution occurs because aquifer sediments contain organic matter that generates anaerobic conditions in the aquifer. These conditions result in the microbial dissolution of iron oxides in the sediment and, thus, the release of the arsenic, normally strongly bound to iron oxides, into the water. As a consequence, arsenic-rich groundwater is often iron-rich, although secondary processes often obscure the association of dissolved arsenic and dissolved iron.. Arsenic is found in groundwater most commonly as the reduced species arsenite and the oxidized species arsenate, being the acute toxicity of arsenite somewhat greater than that of arsenate. Investigations by WHO indicated that 20% of 25,000 boreholes tested in Bangladesh had arsenic concentrations exceeding 50 µg/l.

The occurrence of fluoride is close related to the abundance and solubility of fluoride-containing minerals such as fluorite (CaF_2). Considerably high concentrations of fluoride in groundwater are typically caused by a lack of calcium in the aquifer. Health problems associated with dental fluorosis may occur when fluoride concentrations in groundwater exceed 1.5 mg/l, which is the WHO guideline value since 1984.

The Swiss Federal Institute of Aquatic Science and Technology (EAWAG) has recently developed the interactive Groundwater Assessment Platform (GAP), where the geogenic risk of contamination in a given area can be estimated using geological, topographical and other environmental data without having to test samples from every single groundwater resource. This tool also allows the user to produce probability risk mapping for both arsenic and fluoride.

High concentrations of parameters like salinity, iron, manganese, uranium, radon and chromium, in groundwater, may also be of geogenic origin. This contaminants can be important locally but they are not as widespread as arsenic and fluoride.

On-site Sanitation Systems

A traditional housing compound near Herat, Afghanistan, where a shallow water supply well (foreground) is in close proximity to the pit latrine (behind the white greenhouse) leading to contamination of the groundwater.

Groundwater pollution with pathogens and nitrate can also occur from the liquids infiltrating into the ground from on-site sanitation systems such as pit latrines and septic tanks, depending on the population density and the hydrogeological conditions.

Factors controlling the fate and transport of pathogens are quite complex and the interaction among them is not well understood. If the local hydrogeological conditions (which can vary within a space of a few square kilometres) are ignored, simple on-site sanitation infrastructures such as pit latrines can cause significant public health risks via contaminated groundwater.

Liquids leach from the pit and pass the unsaturated soil zone (which is not completely filled with water). Subsequently, these liquids from the pit enter the groundwater where they may lead to groundwater pollution. This is a problem if a nearby water well is used to supply groundwater for drinking water purposes. During the passage in the soil, pathogens can die off or be adsorbed significantly, mostly depending on the travel time between the pit and the well. Most, but not all pathogens die within 50 days of travel through the subsurface.

The degree of pathogen removal strongly varies with soil type, aquifer type, distance and other environmental factors. For example, the unsaturated zone becomes "washed" during extended periods of heavy rain, providing hydraulic pathway for the quick pass of pathogens. It is difficult to estimate the safe distance between a pit latrine or a septic tank and a water source. In any case, such recommendations about the safe distance are mostly ignored by those building pit latrines. In addition, household plots are of a limited size and therefore pit latrines are often built much closer to groundwater wells than what can be regarded as safe. This results in groundwater pollution and household members falling sick when using this groundwater as a source of drinking water.

Sewage (Treated and Untreated)

Groundwater pollution can be caused by untreated waste discharge leading to diseases like skin lesions, bloody diarrhea and dermatitis. This is more common in locations having limited wastewater treatment infrastructure, or where there are systematic failures of the on-site sewage disposal system. Along with pathogens and nutrients, untreated sewage can also have an important load of heavy metals that may seep into the groundwater system.

The treated effluent from sewage treatment plants may also reach the aquifer if the effluent is infiltrated or discharged to local surface water bodies. Therefore, those substances that are not removed in conventional sewage treatment plants may reach the groundwater as well. For example, detected concentrations of pharmaceutical residues in groundwater were in the order of 50 ng/L in several locations in Germany. This is because in conventional sewage treatment plants, micro-pollutants such as hormones, pharmaceutical residues and other micro-pollutants contained in urine and feces are only partially removed and the remainder is discharged into surface water, from where it may also reach the groundwater.

Groundwater pollution can also occur from leaking sewers which has been observed for example in Germany. This can also lead to potential cross-contamination of drinking-water supplies.

Spreading wastewater or sewage sludge in agriculture may also be included as sources of faecal contamination in groundwater.

Fertilizers and Pesticides

Nitrate can also enter the groundwater via excessive use of fertilizers, including manure spreading. This is because only a fraction of the nitrogen-based fertilizers is converted to produce and other plant matter. The remainder accumulates in the soil or lost as run-off. High application rates of nitrogen-containing fertilizers combined with the high water-solubility of nitrate leads to increased runoff into surface water as well as leaching into groundwater, thereby causing groundwater pollution. The excessive use of nitrogen-containing fertilizers (be they synthetic or natural) is particularly damaging, as much of the nitrogen that is not taken up by plants is transformed into nitrate which is easily leached.

Poor management practices in manure spreading can introduce both pathogens and nutrients (nitrate) in the groundwater system.

The nutrients, especially nitrates, in fertilizers can cause problems for natural habitats and for human health if they are washed off soil into watercourses or leached through soil into groundwater. The heavy use of nitrogenous fertilizers in cropping systems is the largest contributor to anthropogenic nitrogen in groundwater worldwide.

Feedlots/animal corrals can also lead to the potential leaching of nitrogen and metals to groundwater. Over application of animal manure may also result in groundwater pollution with pharmaceutical residues derived from veterinary drugs.

The US Environmental Protection Agency (EPA) and the European Commission are seriously dealing with the nitrate problem related to agricultural development, as a major water supply problem that requires appropriate management and governance.

Runoff of pesticides may leach into groundwater causing human health problems from contaminated water wells. Pesticide concentrations found in groundwater are typically low, and often the regulatory human health-based limits exceeded are also very low. The organophosphorus insecticide monocrotophos (MCP) appears to be one of a few hazardous, persistent, soluble and mobile (it does not bind with minerals in soils) pesticides able to reach a drinking-water source. In general, more pesticide compounds are being detected as groundwater quality monitoring programs have become more extensive; however, much less monitoring has been conducted in developing countries due to the high analysis costs.

Commercial and Industrial Leaks

A wide variety of both inorganic and organic pollutants have been found in aquifers underlying commercial and industrial activities.

Ore mining and metal processing facilities are the primary responsible of the presence of metals in groundwater of anthropogenic origin, including arsenic. The low pH associated with acid mine drainage (AMD) contributes to the solubility of potential toxic metals that can eventually enter the groundwater system.

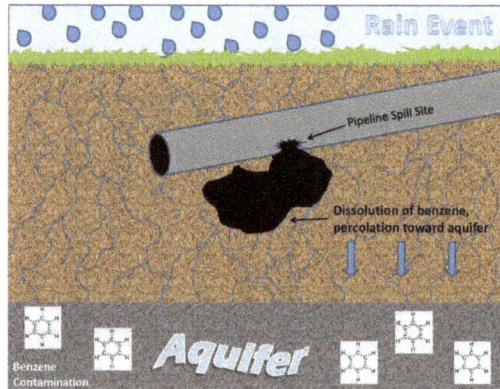

Oil spills associated with underground pipelines and tanks can release benzene and other soluble petroleum hydrocarbons that rapidly percolate down into the aquifer.

There is an increasing concern over the groundwater pollution by gasoline leaked from petroleum underground storage tanks (USTs) of gas stations. BTEX compounds are the most common additives of the gasoline. BTEX compounds, including benzene, have densities lower than water (1 g/ml). Similar to the oil spills on the sea, the non-miscible phase, referred to as Light Non-Aqueous Phase Liquid (LNAPL), will "float" upon the water table in the aquifer.

Chlorinated solvents are used in nearly any industrial practice where degreasing removers are required. PCE is a highly utilized solvent in the dry cleaning industry because of its cleaning effectiveness and relatively low cost. It has also been used for metal-degreasing operations. Because it is highly volatile, it is more frequently found in groundwater than in surface water. TCE has historically been used as a metal cleaning. The military facility Anniston Army Dept (ANAD) in the United States was placed on the US EPA Superfund National Priorities List (NPL) because of groundwater contamination with as much as 27 million pounds of TCE. Both PCE and TCE may degrade to vinyl chloride (VC), the most toxic chlorinated hydrocarbon.

Many types of solvents may have also been disposed illegally, leaking over time to the groundwater system.

Chlorinated solvents such as PCE and TCE have densities higher than water and the non-miscible phase is referred to as Dense Non-Aqueous Phase Liquids (DNAPL). Once they reach the aquifer, they will "sink" and eventually accumulate on the top of low-permeability layers.

Historically, wood-treating facilities have also release insecticides such as pentachlorophenol (PCP) and creosote into the environment, impacting the groundwater resources. PCP is a highly soluble and toxic obsolete pesticide recently listed in the Stockholm Convention on Persistent Organic Pollutants. PAHs and other semi-VOCs are the common contaminants associated with creosote.

Although non-miscible, both LNAPLs and DNAPLs still have the potential to slowly dissolve into the aqueous (miscible) phase to create a plume and thus become a long-term source of contamination. DNAPLs (chlorinated solvents, heavy PAHs, creosote, PCBs) tend to be difficult to manage as they can reside very deep in the groundwater system.

Hydraulic Fracturing

The recent growth of Hydraulic Fracturing ("Fracking") wells in the United States has raised concerns regarding its potential risks of contaminating groundwater resources. The Environmental Protection Agency (EPA), along with many other researchers, has been delegated to study the relationship between hydraulic fracturing and drinking water resources. While it is possible to perform hydraulic fracturing without having a relevant impact on groundwater resources if stringent controls und quality management measures are in place, there are a number of cases where groundwater pollution due to improper handling or technical failures was observed.

While the EPA has not found significant evidence of a widespread, systematic impact on drinking water by hydraulic fracturing, this may be due to insufficient systematic pre- and post- hydraulic fracturing data on drinking water quality, and the presence of other agents of contamination that preclude the link between shale oil/gas extraction and its impact.

Despite the EPA's lack of profound widespread evidence, other researchers have made significant observations of rising groundwater contamination in close proximity to major shale oil/gas drilling sites located in Marcellus*Ellsworth, William (2013). "Injection-Induced Earthquakes". Science AAAS.*</ref> (British Columbia, Canada). Within one kilometer of these specific sites, a subset of shallow drinking water consistently showed higher concentration levels of methane, ethane, and propane concentrations than normal. An evaluation of higher Helium and other noble gas concentration along with the rise of hydrocarbon levels supports the distinction between hydraulic fracturing fugitive gas and naturally occurring "background" hydrocarbon content. This contamination is speculated to be the result of leaky, failing, or improperly installed gas well casings.

Furthermore, it is theorized that contamination could also result from the capillary migration of deep residual hyper-saline water and hydraulic fracturing fluid, slowly flowing through faults and fractures until finally making contact with groundwater resources; however, many researchers argue that the permeability of rocks overlying shale formations are too low to allow this to ever happen sufficiently. To ultimately prove this theory, there would have to be traces of toxic trihalomethanes (THM) since they are often associated with the presence of stray gas contamination, and typically co-occur with high halogen concentrations in hyper-saline waters. Besides, highly saline waters are a common natural feature in deep groundwater systems.

While conclusions regarding groundwater pollution as the result to hydraulic fracturing fluid flow is restricted in both space and time, researchers have hypothesized that the potential for systematic stray gas contamination depends mainly on the integrity of the shale oil/gas well structure, along with its relative geological location to local fracture systems that could potentially provide flow paths for fugitive gas migration.

Though widespread, systematic contamination by hydraulic fracturing has been heavily disputed, one major source of contamination that has the most consensus among researchers of being the most problematic is site-specific accidental spillage of hydraulic fracturing fluid and produced water. So far, a significant majority of groundwater contamination events are derived from surface-level anthropogenic routes rather than the subsurface flow from underlying shale formations. Examples of such events include: a fracking fluid spillage in Acorn Fork Creek, Kentucky that caused a widespread death among aquatic species in 2007; a 420,000 gallon spillage of hyper-saline produced water that turned a once very-fertile farmland in New Mexico into a dead-zone in 2010; and a 42,000 gallon fracking fluid spillage in Arlington, Texas that necessitated an evacuation of over a 100 homes in 2015. While the damage can be obvious, and much more effort is being done to prevent these accidents from occurring so frequently, the lack of data from fracking oil spills continue to leave researchers in the dark. In many of these events, the data acquired from the leakage or spillage is often very vague, and thus would lead researchers to lacking conclusions.

Researchers from the Federal Institute for Geosciences and Natural Resources (BGR) conducted a modelling study for a deep shale-gas formation in the North German Basin. They concluded that the probability is small that the rise of fracking fluids through the geological underground to the surface will impact shallow groundwater.

Landfill Leachate

Leachate from sanitary landfills can lead to groundwater pollution.

Love Canal was one of the most widely known examples of groundwater pollution. In 1978, residents of the Love Canal neighborhood in upstate New York noticed high rates of cancer and an alarming number of birth defects. This was eventually traced to organic solvents and dioxins from an industrial landfill that the neighborhood had been built over and around, which had then infiltrated into the water supply and evaporated in basements to further contaminate the air. Eight hundred families were reimbursed for their homes and moved, after extensive legal battles and media coverage.

Other

Further causes of groundwater pollution are chemical spills from commercial or industrial operations, chemical spills occurring during transport (e.g. spillage of diesel fuels), illegal waste dumping, infiltration from urban runoff or mining operations, road salts, de-icing chemicals from airports and even atmospheric contaminants since groundwater is part of the hydrologic cycle.

The burial of corpses and their subsequent degradation may also pose a risk of pollution to groundwater.

Mechanisms

The passage of water through the subsurface can provide a reliable natural barrier to contamination but it only works under favorable conditions.

The stratigraphy of the area plays an important role in the transport of pollutants. An area can have layers of sandy soil, fractured bedrock, clay, or hardpan. Areas of karst topography on limestone bedrock are sometimes vulnerable to surface pollution from groundwater. Earthquake faults can also be entry routes for downward contaminant entry. Water table conditions are of great importance for drinking water supplies, agricultural irrigation, waste disposal (including nuclear waste), wildlife habitat, and other ecological issues.

Interactions with Surface Water

Although interrelated, surface water and groundwater have often been studied and managed as separate resources. Surface water seeps through the soil and becomes groundwater. Conversely, groundwater can also feed surface water sources. Sources of surface water pollution are generally grouped into two categories based on their origin.

Interactions between groundwater and surface water are complex. Consequently, groundwater pollution, sometimes referred to as groundwater contamination, is not as easily classified as surface water pollution. By its very nature, groundwater aquifers are susceptible to contamination from sources that may not directly affect surface water bodies, and the distinction of point vs. non-point source may be irrelevant. A spill or ongoing release of chemical or radionuclide contaminants into soil (located away from a surface water body) may not create point or non-point source pollution but can contaminate the aquifer below, creating a toxic plume.

Prevention

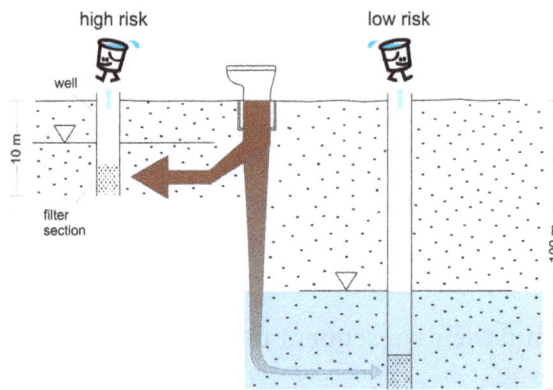

Schematic showing that there is a lower risk of groundwater pollution with greater depth of the water well

Precautionary Principle

The precautionary principle, evolved from Principle 15 of the Rio Declaration on Environment and Development, is important in protecting groundwater resources from pollution. The precautionary principle provides that *"where there are threats of irreversible damage, lack of full scientific certainty shall not be used as reason for postponing cost-effective measures to prevent environmental degradation."*.

One of the six basic principles of the European Union (EU) water policy is the application of the precautionary principle.

Groundwater Quality Monitoring

Groundwater quality monitoring programs have been implemented regularly in many countries around the world. They are important components to understand the hydrogeological system, and for the development of conceptual models and aquifer vulnerability maps.

Groundwater quality must be regularly monitored across the aquifer to determine trends. Effective groundwater monitoring should be driven by a specific objective, for example, a specific contaminant of concern. Contaminant levels can be compared to the World Health Organization (WHO) guidelines for drinking-water quality. It is not rare that limits of contaminants are reduced as more medical experience is gained.

Sufficient investment should be given to continue monitoring over the long term. When a problem is found, action should be taken to correct it. Waterborne outbreaks in the United States decreased with the introduction of more stringent monitoring (and treatment) requirements in the early 90s.

The community can also help monitor the groundwater quality.

Land Zoning for Groundwater Protection

The development of land-use zoning maps has been implemented by several water authorities at different scales around the world. There are two types of zoning maps: aquifer vulnerability maps and source protection maps.

Aquifer Vulnerability Map

It refers to the intrinsic (or natural) vulnerability of a groundwater system to pollution. Intrinsically, some aquifers are more vulnerable to pollution than other aquifers. Shallow unconfined aquifers are more at risk of pollution because there are fewer layers to filter out contaminants.

The unsaturated zone can play an important role in retarding (and in some cases eliminating) pathogens and so must be considered when assessing aquifer vulnerability. The biological activity is greatest in the top soil layers where the attenuation of pathogens is generally most effective.

Preparation of the vulnerability maps typically involves overlaying several thematic maps of physical factors that have been selected to describe the aquifer vulnerability. The index-based parametric mapping method GOD developed by Foster and Hirata (1988) uses three generally available or readily estimated parameters, the degree of Groundwater hydraulic confinement, geological nature of the Overlying strata and Depth to groundwater. A further approach developed by the US EPA named DRASTIC employs seven hydrogeological factors to develop an index of vulnerability: Depth to water table, net Recharge, Aquifer media, Soil media, Topography (slope), Impact on the vadose zone, and hydraulic Conductivity.

There is a particular debate among hydrogeologist whether aquifer vulnerability should be established in a general (intrinsic) way for all contaminants, or specifically for each pollutant.

Source Protection Map

It refers to the capture areas around an individual groundwater source, such as a water well or a spring, to especially protect them from pollution. Thus, potential sources of degradable pollutants, such as pathogens, can be located at distances which travel times along the flowpaths are long enough for the pollutant to be eliminated through filtration or adsorption.

Analytical methods using equations to define groundwater flow and contaminant transport are the most widely used. The WHPA is a semi-analytical groundwater flow simulation program developed by the US EPA for delineating capture zones in a wellhead protection area. The simplest form of zoning employs fixed-distance methods where activities are excluded within a uniformly applied specified distance around abstraction points.

Locating on-site Sanitation Systems

As the health effects of most toxic chemicals arise after prolonged exposure, risk to health from chemicals is generally lower than that from pathogens. Thus, the quality of the source protection measures is an important component in controlling whether pathogens may be present in the final drinking-water.

On-site sanitation systems can be designed in such a way that groundwater pollution from these sanitation systems is prevented from occurring. Detailed guidelines have been developed to estimate safe distances to protect groundwater sources from pollution from on-site sanitation. The following criteria have been proposed for safe siting (i.e. deciding on the location) of on-site sanitation systems:

- Horizontal distance between the drinking water source and the sanitation system

    o Guideline values for horizontal separation distances between on-site sanitation systems and water sources vary widely (e.g. 15 to 100 m horizontal distance between pit latrine and groundwater wells)

- Vertical distance between drinking water well and sanitation system

- Aquifer type

- Groundwater flow direction

- Impermeable layers

- Slope and surface drainage

- Volume of leaking wastewater

- Superposition, i.e. the need to consider a larger planning area

As a very general guideline it is recommended that the bottom of the pit should be at least 2 m above groundwater level, and a minimum horizontal distance of 30 m between a pit and a water source is normally recommended to limit exposure to microbial contamination. However, no general statement should be made regarding the minimum lateral separation distances required to prevent contamination of a well from a pit latrine. For example, even 50 m lateral separation distance might not be sufficient in a strongly karstified system with a downgradient supply well or spring, while 10 m lateral separation distance is completely sufficient if there is a well developed clay cover layer and the annular space of the groundwater well is well sealed.

Legislation

Institutional and legal issues are critical in determining the success or failure of groundwater protection policies and strategies.

Sign near Mannheim, Germany indicating a zone as a dedicated "groundwater protection zone"

United States

In November 2006, the Environmental Protection Agency published the Ground Water Rule in the United States Federal Register. The EPA was worried that the ground water system would be vulnerable to contamination from fecal matter. The point of the rule was to keep microbial pathogens out of public water sources. The 2006 Ground Water Rule was an amendment of the 1996 Safe Drinking Water Act.

The ways to deal with groundwater pollution that has already occurred can be grouped into the following categories: containing the pollutants to prevent them from migrating further; removing the pollutants from the aquifer; remediating the aquifer by either immobilizing or detoxifying the contaminants while they are still in the aquifer (in-si-tu); treating the groundwater at its point of use; or abandoning the use of this aquifer's groundwater and finding an alternative source of water.

Management

Point-of-use Treatment

Portable water purification devices or "point-of-use" (POU) water treatment systems and field water disinfection techniques can be used to remove some forms of groundwater pollution prior to drinking, namely any fecal pollution. Many commercial portable water purification systems or chemical additives are available which can remove pathogens, chlorine, bad taste, odors, and heavy metals like lead and mercury.

Techniques include boiling, filtration, activated charcoal absorption, chemical disinfection, ultraviolet purification, ozone water disinfection, solar water disinfection, solar distillation, homemade water filters.

Arsenic removal filters (ARF) are dedicated technologies typically installed to remove arsenic. Many of these technologies require a capital investment and long-term maintenance. Filters in Bangladesh are usually abandoned by the users due to their high cost and complicated maintenance, which is also quite expensive.

Groundwater Remediation

Groundwater pollution is much more difficult to abate than surface pollution because groundwater can move great distances through unseen aquifers. Non-porous aquifers such as clays partially purify water of bacteria by simple filtration (adsorption and absorption), dilution, and, in some cases, chemical reactions and biological activity; however, in some cases, the pollutants merely transform to soil contaminants. Groundwater that moves through open fractures and caverns is not filtered and can be transported as easily as surface water. In fact, this can be aggravated by the human tendency to use natural sinkholes as dumps in areas of karst topography.

Pollutants and contaminants can be removed from ground water by applying various techniques thereby making it safe for use. Ground water treatment (or remediation) techniques span biological, chemical, and physical treatment technologies. Most ground water treatment techniques utilize a combination of technologies. Some of the biological treatment techniques include bioaugmentation, bioventing, biosparging, bioslurping, and phytoremediation. Some chemical treatment techniques include ozone and oxygen gas injection, chemical precipitation, membrane separation, ion exchange, carbon absorption, aqueous chemical oxidation, and surfactant enhanced recovery. Some chemical techniques may be implemented using nanomaterials. Physical treatment techniques include, but are not limited to, pump and treat, air sparging, and dual phase extraction.

Abandonment

If treatment or remediation of the polluted groundwater is deemed to be too difficult or expensive then abandoning the use of this aquifer's groundwater and finding an alternative source of water is the only other option.

Society and Culture

Examples

Hinkley, U.S.

The town of Hinkley, California (U.S.), had its groundwater contaminated with hexavalent chromium starting in 1952, resulting in a legal case against Pacific Gas & Electric (PG&E) and a multimillion-dollar settlement in 1996. The legal case was dramatized in the film *Erin Brockovich*, released in 2000.

California, U.S.

Nitrates and water contamination in California's Central Valley

Walkerton, Canada

In the year 2000, groundwater pollution occurred in the small town of Walkerton, Canada leading to seven deaths in what is known as the Walkerton *E. Coli* outbreak. The water supply which was drawn from groundwater became contaminated with the highly dangerous O157:H7 strain of *E. coli* bacteria. This contamination was due to farm runoff into an adjacent water well that was vulnerable to groundwater pollution.

Lusaka, Zambia

The peri-urban areas of Lusaka, the capital of Zambia, have ground conditions which are strongly karstified and for this reason – together with the increasing population density in these peri-urban areas – pollution of water wells from pit latrines is a major public health threat there.

Concept of Groundwater flow

Consider the flow of ground water taking place within a small cube (of lengths Δx, Δy and Δz respectively the direction of the three areas which may also be called the elementary control volume) of a saturated aquifer as shown in Figure a.

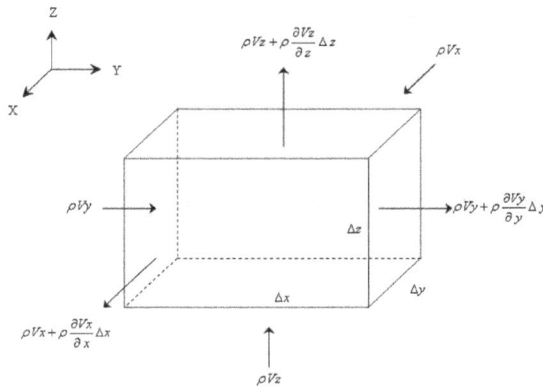

FIGURE a. Infinitesimal cube for deriving the equation of continuity of flow of ground water

It is assumed that the density of water (ρ) does not change in space along the three directions which implies that water is considered incompressible. The velocity components in the x, y and z directions have been denoted as v_x, v_y, v_z respectively.

Since water has been considered incompressible, the total incoming water in the cuboidal volume should be equal to that going out. Defining inflows and outflows as:

Inflows:

In x-direction: $\rho\, v_x\, (\Delta y.\Delta x)$

In y-direction: $\rho\, v_y\, (\Delta x.\Delta z)$

In z-direction: $\rho\, v_z\, (\Delta x.\Delta y)$

Outflows:

$$\text{In X-direction: } \rho\left(v_x + \frac{\partial vx}{\partial x}\Delta x \Delta x\right)(\Delta y.\Delta z)$$

$$\text{In Y-direction: } \rho\left(v_y + \frac{\partial vy}{\partial y}\Delta y\right)(\Delta x.\Delta z)$$

$$\text{In Z-direction: } \rho\left(v_z + \frac{\partial vz}{\partial z}\Delta z\right)(\Delta y.\Delta x)$$

The net mass flow per unit time through the cube works out to:

$$\left[\frac{\partial v_x}{\partial x} + \frac{\partial v_y}{\partial y} + \frac{\partial v_z}{\partial z}\right](\Delta x.\Delta y.\Delta z) \qquad (1)$$

Or

$$\frac{\partial v_x}{\partial x} + \frac{\partial v_y}{\partial y} + \frac{\partial v_z}{\partial z} = 0 \qquad (2)$$

This is continuity equation for flow. But this water flow is due to a difference in potentiometric head per unit length in the direction of flow. A relation between the velocity and potentiometric gradient was first suggested by Henry Darcy, a French Engineer, in the mid nineteenth century. He found experimentally that the discharge 'Q' passing through a tube of cross sectional area 'A' filled with a porous material is proportional to the difference of the hydraulic head 'h' between the two end points and inversely proportional to the flow length 'L'.

It may be noted that the total energy (also called head, h) at any point in the ground water flow per unit weight is given as

$$h = Z + \frac{p}{\gamma} + \frac{v^2}{2g} \qquad (3)$$

Where

Z is the elevation of the point above a chosen datum;

$\frac{p}{\gamma}$ is the pressure head, and

$\frac{v^2}{2g}$ is the velocity head

Since the ground water flow velocities are usually very small, $\frac{v^2}{2g}$ is neglected and

$h = Z + \frac{p}{\gamma}$ is termed as the potentiometric head (or piezometric head in some texts)

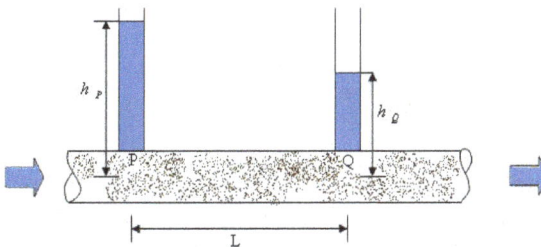

FIGURE b. Flow through a saturated porous medium

Thus

$$Q \alpha A \cdot \left(\frac{h_P - h_Q}{L} \right) \qquad (4)$$

Introducing proportionality constant K, the expression becomes

$$Q = K.A. \left(\frac{h_P - h_Q}{L} \right) \tag{5}$$

Since the hydraulic head decreases in the direction of flow, a corresponding differential equation would be written as

$$Q = -K.A. \left(\frac{dh}{dl} \right) \tag{6}$$

Where (dh/dl) is known as hydraulic gradient.

The coefficient 'K' has dimensions of L/T, or velocity.

Thus the velocity of fluid flow would be:

$$v = \frac{Q}{A} = -K \left(\frac{dh}{dl} \right) \tag{7}$$

It may be noted that this velocity is not quite the same as the velocity of water flowing through an open pipe. In an open pipe, the entire cross section of the pipe conveys water. On the other hand, if the pipe is filed with a porous material, say sand, then the water can only flow through the pores of the sand particles. Hence, the velocity obtained by the above expression is only an apparent velocity, with the actual velocity of the fluid particles through the voids of the porous material is many time more. But for our analysis of substituting the expression for velocity in the three directions x, y and z in the continuity relation, equation (2) and considering each velocity term to be proportional to the hydraulic gradient in the corresponding direction, one obtains the following relation

$$\frac{\partial}{\partial x}\left(K_x \frac{\partial h}{\partial x} \right) + \frac{\partial}{\partial x}\left(K_y \frac{\partial h}{\partial y} \right) + \frac{\partial}{\partial z}\left(K_z \frac{\partial h}{\partial z} \right) = 0 \tag{8}$$

Here, the hydraulic conductivities in the three directions (K_x, K_y and K_z) have been assumed to be different as for a general anisotropic medium. Considering isotropic medium with a constant hydraulic conductivity in all directions, the continuity equation simplifies to the following expression:

$$\frac{\partial^2 h}{\partial x^2} + \frac{\partial^2 h}{\partial y^2} + \frac{\partial^2 h}{\partial z^2} = 0 \tag{9}$$

In the above equation, it is assumed that the hydraulic head is not changing with time, that is, a steady state is prevailing.

If now it is assumed that the potentiometric head changes with time at the location of the control volume, then there would be a corresponding change in the porosity of the aquifer even if the fluid density is assumed to be unchanged.

Important term:

Porosity: It is ratio of volume of voids to the total volume of the soil and is generally expressed as percentage.

Ground Water Flow Equations Under Unsteady State

For an unsteady case, the rate of mass flow in the elementary control volume is given by:

$$\rho\left[\frac{\partial v_x}{\partial x}+\frac{\partial v_y}{\partial y}+\frac{\partial v_z}{\partial z}\right]\Delta x\Delta y\Delta z=\frac{\partial M}{\partial t} \qquad (10)$$

This is caused by a change in the hydraulic head with time plus the porosity of the media increasing accommodating more water. Denoting porosity by the term 'n', a change in mass 'M' of water contained with respect to time is given by

$$\frac{\partial M}{\partial t}=\frac{\partial}{\partial t}\left(\rho n\Delta x\Delta y\Delta z\right) \qquad (11)$$

Considering no lateral strain, that is, no change in the plan area $\Delta x.\Delta y$ of the control volume, the above expression may be written as:

$$\frac{\partial M}{\partial t}=\frac{\partial \rho}{\partial t}\left(n\Delta x\Delta y\Delta z\right)+\frac{\partial}{\partial t}\left(n.\Delta z\right).\rho\Delta x\Delta y \qquad (12)$$

Where the density of water (ρ) is assumed to change with time. Its relation to a change in volume of the water V_w, contained within the void is given as:

$$\frac{d\left(V_w\right)}{V_w}=-\frac{d\rho}{\rho} \qquad (13)$$

The negative sign indicates that a reduction in volume would mean an increase in the density from the corresponding original values.

The compressibility of water, β, is defined as:

$$\beta=-\frac{\left[\dfrac{d\left(V_w\right)}{V_w}\right]}{dp} \qquad (14)$$

Where 'dp' is the change in the hydraulic head 'p' Thus,

$$\beta = \frac{d\rho}{\rho dp} \qquad (15)$$

That is,

$$d\rho = \rho \, dp \, \beta \qquad (16)$$

The compressibility of the soil matrix, α, is defined as the inverse of E_s, the elasticity of the soil matrix. Hence

$$\frac{1}{\alpha} = E_s = -\frac{d(\sigma_z)}{\dfrac{d(\Delta z)}{\Delta z}} \qquad (17)$$

Where σ_z is the stress in the grains of the soil matrix.

Now, the pressure of the fluid in the voids, p, and the stress on the solid particles, σ_z must combine to support the total mass lying vertically above the elementary volume. Thus,

$$p + \sigma_z = \text{constant} \qquad (18)$$

Leading to

$$d\sigma_z = -dp \qquad (19)$$

Thus,

$$\frac{1}{\alpha} = -\frac{dp}{\dfrac{d(\Delta z)}{\Delta z}} \qquad (20)$$

Also since the potentiometric head 'h' given by

$$h = \frac{p}{\gamma} + Z \qquad (21)$$

Where Z is the elevation of the cube considered above a datum. We may therefore re-write the above as

$$\frac{dh}{dz} = \frac{1}{\gamma}\frac{dp}{dz} + 1 \qquad (22)$$

First term for the change in mass 'M' of the water contained in the elementary volume, Equation 12, is

$$\frac{dp}{dt}.n.\Delta x\Delta y\Delta z \qquad (23)$$

This may be written, based on the derivations shown earlier, as equal to

$$n.\rho.\beta.\frac{\partial p}{\partial t}.\Delta x\Delta y\Delta z \qquad (24)$$

Also the volume of soil grains, V_s, is given as

$$V_S = (1-n)\Delta x\Delta y\Delta z \qquad (25)$$

Thus,

$$dV_S = [d(\Delta z)-d(n\Delta z)]\Delta x\Delta y \qquad (26)$$

Considering the compressibility of the soil grains to be nominal compared to that of the water or the change in the porosity, we may assume dVS to be equal to zero. Hence,

$$[d(\Delta z)-d(n\Delta z)]\Delta x\Delta y = 0 \qquad (27)$$

Or,

$$d(\Delta z) = d(n\Delta z) \qquad (28)$$

Which may substituted in second term of the expression for change in mass, M, of the elementary volume, changing it to

$$\frac{\partial(n\Delta z)}{\partial t}\rho\Delta x\Delta y$$

$$=\frac{\partial(\Delta z)}{\partial t}\rho\Delta x\Delta y$$

$$=\rho\frac{\frac{\partial(\Delta z)}{\Delta z}}{\partial t}\Delta x\Delta y\Delta z$$

$$=\rho\alpha\frac{\partial p}{\partial t}\Delta x\Delta y\Delta z \qquad (29)$$

Thus, the equation for change of mass, M, of the elementary cubic volume becomes

$$\frac{\partial M}{\partial t}=(\alpha+n\beta).\rho\frac{\partial p}{\partial t}\Delta x\Delta y\Delta z \qquad (30)$$

Combining Equation (30) with the continuity expression for mass within the volume, equation (10), gives the following relation:

$$-\rho\left[\frac{\partial v_x}{\partial x}+\frac{\partial v_y}{\partial y}+\frac{\partial v_z}{\partial z}\right]=(\alpha+n\beta)\rho\frac{\partial p}{\partial t} \qquad (31)$$

Assuming isotropic media, that is, $K_x=K_y=K_z=K$ and applying Darcy's law for the velocities in the three directions, the above equation simplifies to

$$K\left[\frac{\partial^2 h}{\partial x^2}+\frac{\partial^2 h}{\partial y^2}+\frac{\partial^2 h}{\partial z^2}\right]=(\alpha+n\beta)\rho\frac{\partial p}{\partial t} \qquad (32)$$

Now, since the potentiometric (or hydraulic) head h is given as

$$h=\frac{p}{\gamma}+Z \qquad (33)$$

The flow equation can be expressed as

$$\left[\frac{\partial^2 h}{\partial x^2}+\frac{\partial^2 h}{\partial y^2}+\frac{\partial^2 h}{\partial z^2}\right]=(\alpha+n\beta)\frac{\gamma}{K}\frac{\partial h}{\partial t} \qquad (34)$$

The above equation is the general expression for the flow in three dimensions for an isotropic homogeneous porous medium. The expression was derived on the basis of an elementary control volume which may be a part of an unconfined or a confined aquifer.

Ground Water Flow Expressions for Ground Water Flow Unconfined And Confined Aquifers

Unsteady flow takes place in an unconfined and confined aquifer would be either due to:

- Change in hydraulic head (for unconfined aquifer) or potentiometric head (for confined aquifer) with time.

- And, or compressibility of the mineral grains of the soil matrix forming the aquifer

- And, or compressibility of the water stored in the voids within the soil matrix

We may visually express the above conditions as shown in Figure c, assuming an increase in the hydraulic (or potentiometric head) and a compression of soil matrix and pore water to accommodate more water

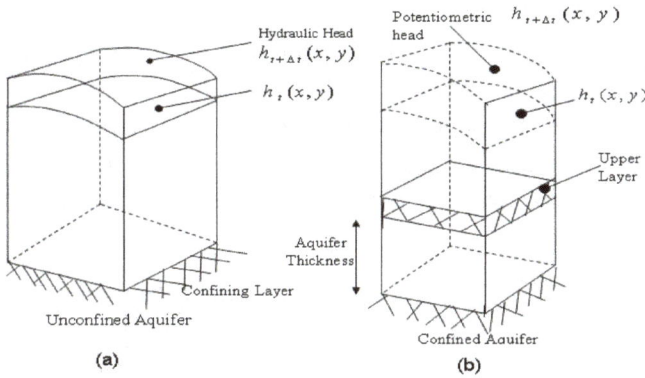

FIGURE c. (a) Free surface of ground water table in unconfined aquifer
(b) Potentiometric surface in confined aquifer

Since storability S of a confined aquifer was defined as

$$S = b\gamma(\alpha + n\beta) \qquad (35)$$

The flow equation for a confined aquifer would simplify to the following:

$$\left[\frac{\partial^2 h}{\partial x^2} + \frac{\partial^2 h}{\partial y^2} + \frac{\partial^2 h}{\partial z^2}\right] = \frac{S}{Kb}\frac{\partial h}{\partial t} \qquad (36)$$

Defining the transmissivity T of a confined aquifer as a product of the hydraulic conductivity K and the saturated thickness of the aquifer, b, which is:

$$T = K \cdot b \qquad (37)$$

The flow equation further reduces to the following for a confined aquifer

$$\left[\frac{\partial^2 h}{\partial x^2} + \frac{\partial^2 h}{\partial y^2} + \frac{\partial^2 h}{\partial z^2}\right] = \frac{S}{T}\frac{\partial h}{\partial t} \qquad (38)$$

For unconfined aquifers, the storability S is given by the following expression

$$S = S_y + hS_s \qquad (39)$$

Where S_y is the specific yield and S_s is the specific storage equal to $\gamma(\alpha + n\beta)$

Usually, S_s is much smaller in magnitude than S_y and may be neglected. Hence S under water table conditions for all practical purposes may be taken equal to S_y.

Two Dimensional Flow in Aquifers

Under many situations, the water table variation (for unconfined flow) in areal extent is not much, which means that there the ground water flow does not have much of a vertical velocity component. Hence, a two – dimensional flow situation may be approximated for these cases. On the other hand, where there is a large variation in the water table under certain situation, a three dimensional velocity field would be the correct representation as there would be significant component of flow in the vertical direction apart from that in the horizontal directions. This difference is shown in the illustrations given in Figure.

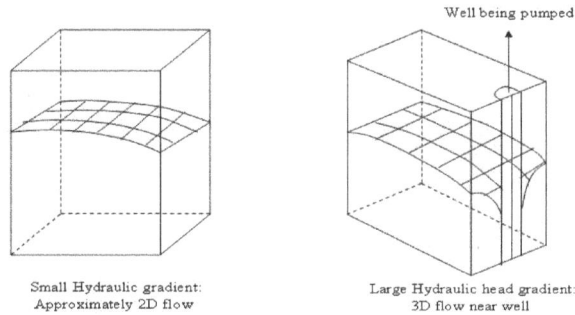

Small Hydraulic gradient:
Approximately 2D flow

Large Hydraulic head gradient:
3D flow near well

Well being pumped

FIGURE d. Difference between small and large hydraulic gradients of ground water table

In case of two dimensional flow, the equation flow for both unconfined and confined aquifers may be written as,

$$\left[\frac{\partial^2 h}{\partial x^2} + \frac{\partial^2 h}{\partial y^2}\right] = \frac{S}{T}\frac{\partial h}{\partial t} \qquad (40)$$

There is one point to be noted for unconfined aquifers for hydraulic head (or water table) variations with time. It is that the change in the saturated thickness of the aquifer with time also changes the transmissivity, T, which is a product of hydraulic conductivity K and the saturated thickness h. The general form of the flow equation for two dimensional unconfined flow is known as the Boussinesq equation and is given as

$$\frac{\partial}{\partial x}\left(h\frac{\partial h}{\partial x}\right) + \frac{\partial}{\partial y}\left(h\frac{\partial h}{\partial y}\right) = \frac{S_y}{K}\frac{\partial h}{\partial t} \qquad (41)$$

Where S_y is the specific yield.

If the drawdown in the unconfined aquifer is very small compared to the saturated thickness, the variable thickness of the saturated zone, h, can be replaced by an average thickness, b, which is assumed to be constant over the aquifer.

For confined aquifer under an unsteady condition though the potentiometric surface declines, the saturated thickness of the aquifer remains constant with time and is equal to an average value 'b'. Solving the ground water flow equations for flow in aquifers require the help of numerical techniques, except for very simple cases.

Two Dimensional Seepage Flow

Another example of two dimensional flow would that be when the flow can be approx-imated. to be taking place in the vertical plane. Such situations might occur as for the seepage taking place below a dam as shown in Figure.

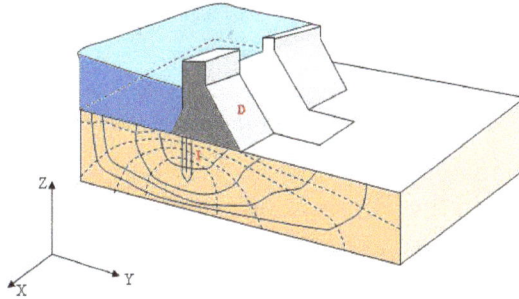

FIGURE e. Seepage flow below a concrete gravity dam

Under steady state conditions, the general equation of flow, considering an isotropic porous medium would be

$$\frac{\partial^2 h}{\partial y^2} + \frac{\partial^2 h}{\partial z^2} = 0 \qquad (42)$$

However, solving the above Equation (42) for would require advanced analytical meth-ods or numerical techniques. More about seepage flow would be discussed in the later session.

Steady One Dimensional Flow in Aquifers

Some simplified cases of ground water flow, usually in the vertical plane, can be ap-proximated by one dimensional equation which can then be solved analytically.

Confined Aquifers

If there is a steady movement of ground water in a confined aquifer, there will be a gra-dient or slope to the potentiometric surface of the aquifer. The gradient, again, would be decreasing in the direction of flow. For flow of this type, Darcy's law may be used directly.

Aquifer with Constant Thickness

This situation may be shown as in Figure f.

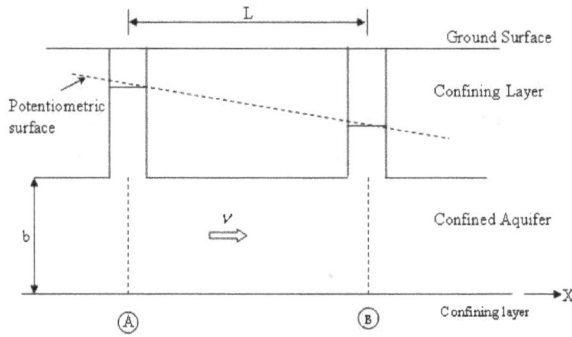

FIGURE F. Flow through an aquifer of constant thckness

Assuming unit thickness in the direction perpendicular to the plane of the paper, the flow rate 'q' (per unit width) would be expressed for an aquifer of thickness 'b'

$$q = b*1*v \qquad (43)$$

According to Darcy's law, the velocity 'v' is given by

$$v = -K \frac{\partial h}{\partial x} \qquad (44)$$

Where h, the potentiometric head, is measured above a convenient datum. Note that the actual value of 'h' is not required, but only its gradient $\frac{\partial h}{\partial x}$ in the direction of flow, x, is what matters. Here is K is the hydraulic conductivity

Hence,

$$q = bK \frac{\partial h}{\partial x} \qquad (45)$$

The partial derivative of 'h' with respect to 'x' may be written as normal derivative since we are assuming no variation of 'h' in the direction normal to the paper. Thus

$$q = -bK \frac{dh}{dx} \qquad (46)$$

For steady flow, q should not vary with time, t, or spatial coordinate, x. hence,

$$\frac{dq}{dx} = -bK \frac{d^2 h}{dx^2} = 0 \qquad (47)$$

Since the width, b, and hydraulic conductivity, K, of the aquifer are assumed to be constants, the above equation simplifies to:

$$\frac{d^2h}{dx^2} = 0 \qquad\qquad (48)$$

Which may be analytically solved as

$$h = C_1 x + C_2 \qquad\qquad (49)$$

Selecting the origin of coordinate x at the location of well A (as shown in Figure f), and having a hydraulic head, h_A and also assuming a hydraulic head of well B, located at a distance L from well A in the x-direction and having a hydraulic head h_B, we have:

$$h_A = C_1.0 + C_2 \text{ and}$$
$$h_B = C_1.L + C_2$$

Giving

$$C_1 = h\text{-}h_A/L \text{ and } C_2 = h_A \qquad\qquad (50)$$

Thus the analytical solution for the hydraulic head 'h' becomes:

$$H = \frac{h_B - h_A}{L} x + h_A \qquad\qquad (51)$$

Aquifer with Variable Thickness

Consider a situation of one- dimensional flow in a confined aquifer whose thickness, b, varies in the direction of flow, x, in a linear fashion as shown in Figure g.

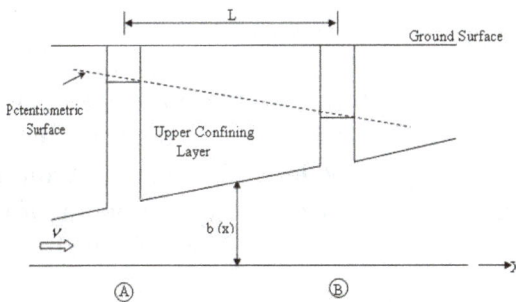

FIGURE g. Flow through an aquifer with variable thickness

The unit discharge, q, is now given as

$$q = -b(x)K\frac{dh}{dx} \qquad\qquad (52)$$

Where K is the hydraulic conductivity and dh/dx is the gradient of the potentiometric surface in the direction of flow, x.

For steady flow, we have,

$$\frac{dq}{dx} = -K\left[\frac{db}{dx}\frac{dh}{dx} + b\frac{d^2h}{dx^2}\right] = 0 \qquad (53)$$

Which may be simplified, denoting $\frac{db}{dx}$ as b'

$$b\frac{d^2h}{dx^2} + b'\frac{dh}{dx} = 0 \qquad (54)$$

A solution of the above differential equation may be found out which may be substituted for known values of the potentiometric heads h_A and h_B in the two observation wells A and B respectively in order to find out the constants of integration.

Unconfined Aquifers

In an unconfined aquifer, the saturated flow thickness, h is the same as the hydraulic head at any location, as seen from Figure h:

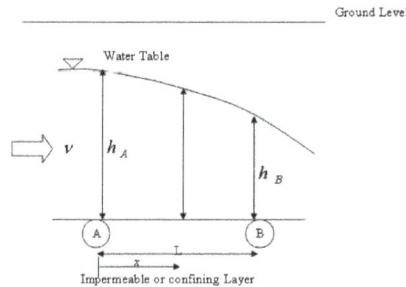

FIGURE h. Flow through an unconfined aquifer

Considering no recharge of water from top, the flow takes place in the direction of fall of the hydraulic head, h, which is a function of the coordinate, x taken in the flow direction. The flow velocity, v, would be lesser at location A and higher at B since the saturated flow thickness decreases. Hence v is also a function of x and increases in the direction of flow. Since, v, according to Darcy's law is shown to be

$$v = K\frac{dh}{dx} \qquad (55)$$

the gradient of potentiometric surface, dh/dx, would (in proportion to the velocities) be smaller at location A and steeper at location B. Hence the gradient of water table in unconfined flow is not constant, it increases in the direction of flow.

This problem was solved by J.Dupuit, a French hydraulician, and published in 1863 and his assumptions for a flow in an unconfined aquifer is used to approximate the flow situation called Dupuit flow. The assumptions made by Dupuit are:

- The hydraulic gradient is equal to the slope of the water table, and

- For small water table gradients, the flow-lines are horizontal and the equi-potential lines are vertical.

The second assumption is illustrated in Figure.

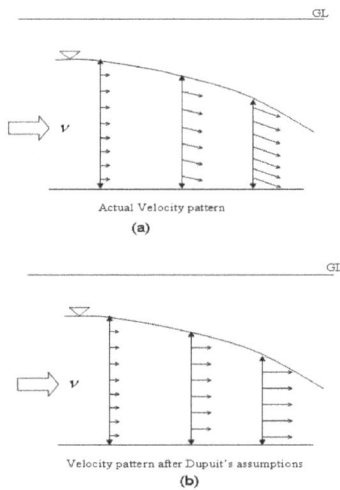

FIGURE i. (a) Actual velocity pattern in ground water flow and
(b) Assumption of Dupuit regarding ground water flow

Solutions based on the Dupuit's assumptions have proved to be very useful in many practical purposes. However, the Dupuit assumption do not allow for a seepage face above an outflow side.

An analytical solution to the flow would be obtained by using the Darcy equation to express the velocity, v, at any point, x, with a corresponding hydraulic gradient $\dfrac{dh}{dx}$ as

$$v = -K \frac{dh}{dx} \tag{56}$$

Thus, the unit discharge, q, is calculated to be

$$q = -Kh \frac{dh}{dx} \tag{57}$$

Considering the origin of the coordinate x at location A where the hydraulic head us h_A and knowing the hydraulic head h_B at a location B, situated at a distance L from A, we may integrate the above differential equation as:

$$\int_{0}^{L} q dx = -K \int_{h_A}^{h_B} h dh \qquad (58)$$

Which, on integration, leads to

$$q x \Big|_{0}^{L} = -K . \frac{h^2}{2} \Big|_{h_A}^{h_B} \qquad (59)$$

Or,

$$q.L=K\left[\frac{h_B^2}{2} - \frac{h_A^2}{2}\right] \qquad (60)$$

Rearrangement of above terms leads to, what is known as the Dupuit equation:

$$q = -\frac{1}{2}K\left[\frac{h_B^2 - h_A^2}{L}\right] \qquad (61)$$

An example of the application of the above equation may be for the ground water flow in a strip of land located between two water bodies with different water surface elevations, as in Figure j.

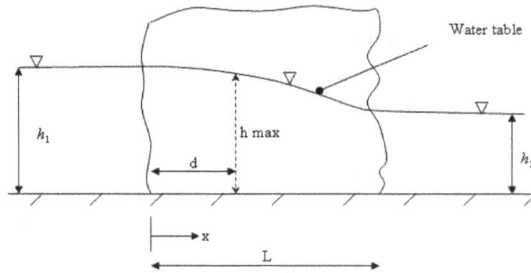

FIGURE j . Ground water flow through a strip of land with difference in water surface elevation on either side.

The equation for the water table, also called the phreatic surface may be derived from Equation (61) as follows:

$$h = \sqrt{h_1^2 - \left(h_1^2 - h_2^2\right)\frac{x}{L}} \qquad (62)$$

In case of recharge due to a constant infiltration of water from above the water table rises to a many as shown in Figure k.

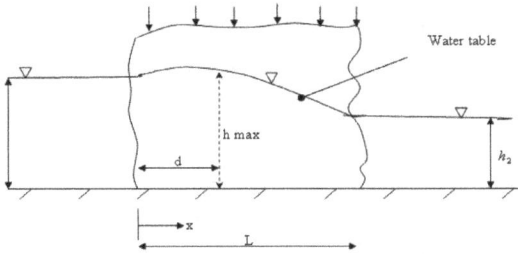

FIGURE k . Ground water flow through a strip of land with infiltration from above

There is a difference with the earlier cases, as the flow per unit width, q, would be increasing in the direction of flow due to addition of water from above. The flow may be analysed by considering a small portion of flow domain as shown in Figure l.

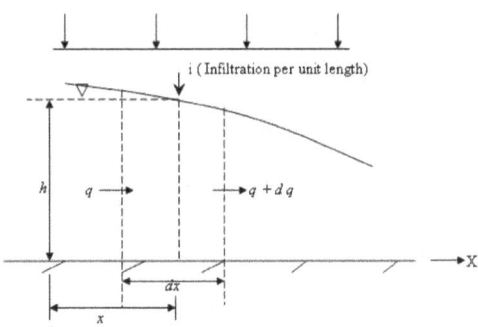

FIGURE l . Definition of terms for flow analysis for the case shown in Figure k

Considering the infiltration of water from above at a rate i per unit length in the direction of ground water flow, the change in unit discharge d_q is seen to be

$$d_q = i.dx \qquad (63)$$

Or,

$$\frac{dq}{dx} = i \qquad (64)$$

From Darcy's law,

$$q = -K.h.\frac{dh}{dx} = -\frac{1}{2} k \frac{d(h^2)}{dx^2} \qquad (65)$$

$$\frac{dq}{dx} = -\frac{1}{2} K \frac{d^2(h^2)}{dx^2} \qquad (66)$$

Substituting the expression for $\dfrac{dq}{dx}$ we have,

$$-\frac{1}{2} K \frac{d^2\left(h^2\right)}{dx^2} = i \tag{67}$$

Or,

$$\frac{d^2\left(h^2\right)}{dx^2} = \frac{2.i}{k} \tag{68}$$

The solution for this equation is of the form

$$Kh^2 + 2x^2 = C_1 x + C_2 \tag{69}$$

If, now, the boundary condition is applied as,

At x = 0, h = h$_1$, and

At x = L, h = h$_2$

The equation for the water table would be:

$$h = \sqrt{h_1^2 - \left(\frac{h_1^2 - h_1^2}{L}\right) x + \frac{i}{K}(L - x)x} \tag{70}$$

And,

$$q_x = q_0 + 2x \tag{71}$$

Where q_0 is the unit discharge at the left boundary, x = 0, and may be found out to be

$$q_0 = \left(\frac{h_1^2 - h_1^2}{2L}\right) - \frac{iL}{2} \tag{72}$$

Which gives an expression for unit discharge q_x at any point x from the origin as

$$q_x = K\left(\frac{h_1^2 - h_1^2}{2L}\right) - i\left(\frac{L}{2} - x\right) \tag{73}$$

For no recharge due to infiltration, i = 0 and the expression for q_x is then seen to become independent of x, hence constant, which is expected.

Flow Dynamics

It is common for water resources engineers to design a water system involving flow of water from one place to another, usually passing a variety of structures on the way some of them meant for controlling the flow quantity. Rivers and artificial channels, like canals, convey water with a free surface, that is, the surface of water being exposed to air as opposed to flow of water in pipes. It is easy to visualize that for any such open channel flow, as they are called; the presence or absence of a hydraulic structure controls the position of the free surface of water. Knowing the mathematical description of flowing water, it is possible to compute the water surface profile, which is important for example in designing the height of the channel walls of the water conveying system.

Another example, the case of river flow obstruction by the presence dam may be mentioned. The water level of the river increases on construction of the dam and it is essential to know the maximum possible rise, perhaps during the maximum flood, in order to know the degree of submergence of the land behind the dam. Barrages are low height structures, and hence, the rise of water will not be occurring uniformly across the river, again due to the difference of gate operation.

In this lesson, the behavior and corresponding mathematical description of flow in open channels are reviewed in order to utilize them in designing water resources systems.'

Flow in Natural Rivers

Figure a below shows a river carrying a low discharge.

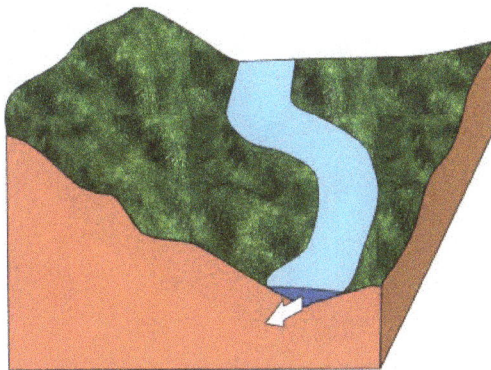

FIGURE. a A river flowing within its banks .

When the water surface of the river just touches its banks, the discharge flowing through the river at this stage is called the "bank full discharge". It is also sometimes called the "dominant discharge". If the discharge in the river increases, the water will overflow the banks and would spill over to the adjacent land, called the flood plains (Figure b).

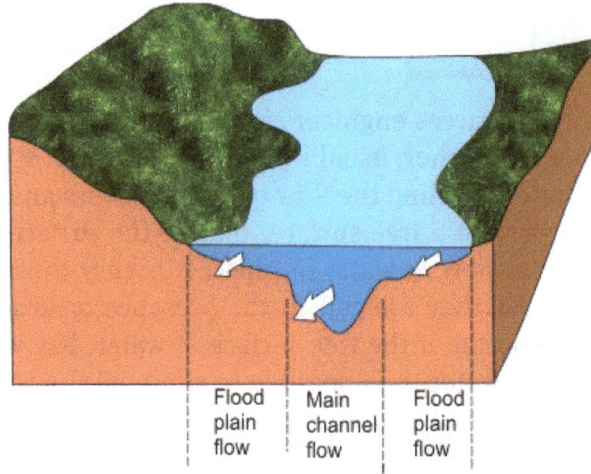

FIGURE .b A river flowing its banks
during flood

Though the amount of discharge flowing through the river is of interest to the water resources engineer it cannot be measured directly by any instruments. Rather, an indirect method is used which requires knowledge of the velocity distribution in a river or an open channel.

If we plot the velocity profile across a river, as shown in Figure a, it would actually vary in three dimensions. Figure c shows the variation of velocity at the water surface.

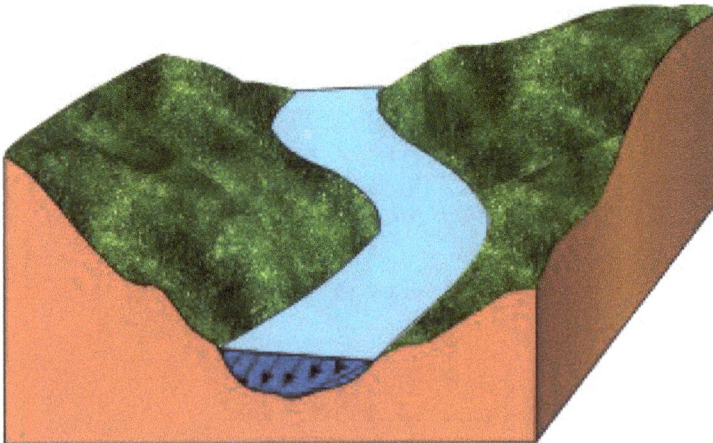

FIGURE C. Variation of surface velocity across a river section

It may be observed that velocity is highest at the center of the river but is zero at the banks. If a velocity profile were plotted on another horizontal plane at certain depth of the river, there too the velocity profile would be found to be similar in shape, but smaller in magnitude (Figure d).

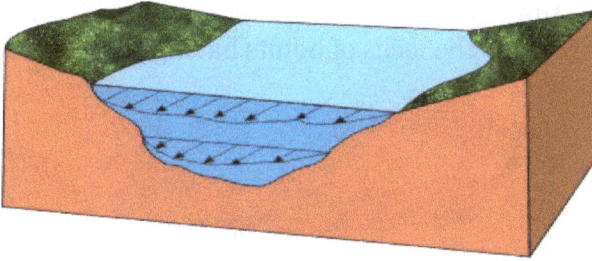

FIGURE d. Velocity variation of a river section at two levels

Similarly the velocity profile of the river flowing in flood would be as shown in Figure e, showing that the velocities over the flood plains is smaller compared to the main stream flow.

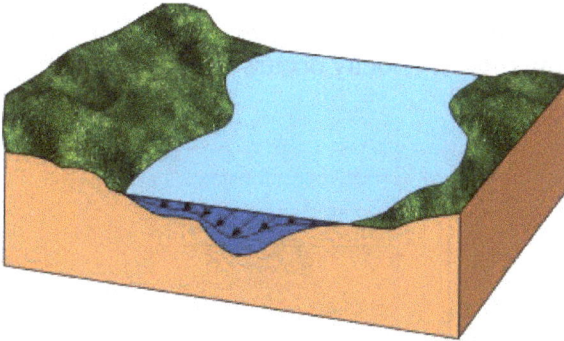

FIGURE e. Surface velocity profile across a river section for a river flowing in flood

If we now take a look at the variation of velocity in a vertical plane within a river, and we plot them along different vertical lines across the river, then we may find the velocity profiles similar to those shown in Figure f.

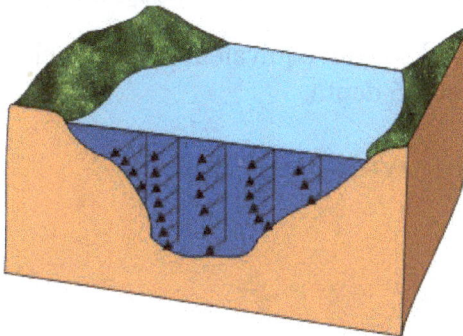

FIGURE f. Vertical velocity profiles in a
river section

In order to measure the discharge being conveyed in a river, the velocity profile or the average velocity at a number of equally spaced sections are measured, as in Figure f. The total discharge is then approximately taken equal to the sum of the discharges passing through each segment.

Another way of depicting the velocity variation across a river cross-section is to plot "Isovels", which are actually the locus of points having equal velocity (Figure g).

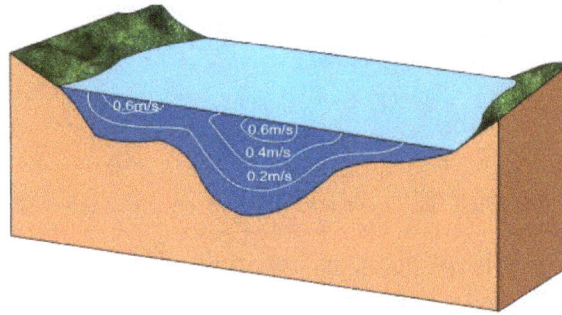

FIGURE g. Isovels showing contours of typical equal velocities across a typical river section

It has been observed through experiments that a plot of velocity in the vertical plane would show that the maximum velocity occurs slightly below the surface (Figure h) for a typical river flow.

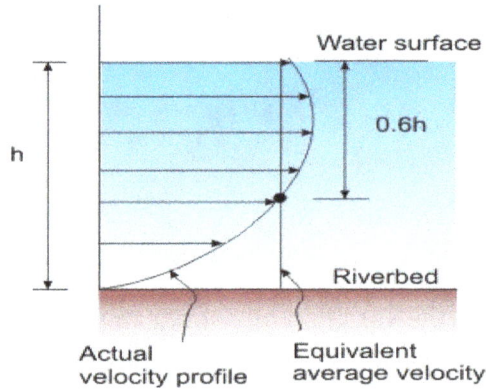

FIGURE h. Vertical velocity distribution and average velocity for flow in a river

It has further been observed that an equivalent average velocity is almost equal to the actual velocity measured at 0.6 depth.

Variation of Discharge with River Stage

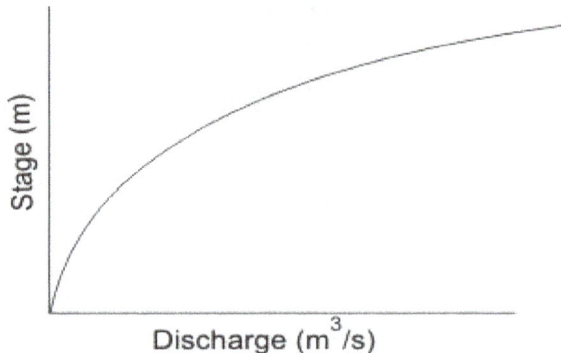

FIGURE i. Stage-discharge curve for a river

The water level in a river is sometimes called the "stage" and as this varies, there is a proportional change in the total discharge conveyed. For each point of a river, the relation between stage and discharge is unique but a general form is found to be as shown in Figure i.

The general mathematical description for the stage-discharge relation is given as:

$$Q = k \left(h - h_0 \right)^m$$

Where h is the gauge corresponding to a discharge Q and ho is the corresponding to zero discharge k and m are constants. If the variables (Q and H) are plotted on a log-log graph, then it generally plots in a straight line as:

$$\log Q = m \log \left(h - h_0 \right) + \log k$$

Flow Variation Along River Length

It may be interpreted from Figures d or f that the velocity in a river cross section actually varies from bank to bank and from riverbed to free water surface and hence, can be called a two dimensional variation in a vertical plane. However, for engineering purposes it is, sufficient, generally, to use an equivalent velocity in the direction of river motion (perpendicular to river cross section) which may be obtained by dividing the total discharge by the cross sectional area. In a natural river, therefore, these flow velocities may vary from section to section (Figure j).

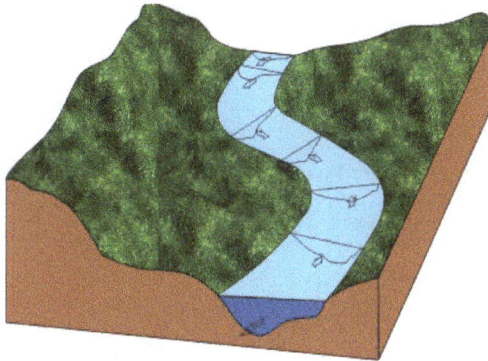

FIGURE j. Variation of cross section and velocity along the length of a river showing vertical section at each point

If we now consider an axis along the length of the river, the total energy (H) is given as:

$$H = Z + h + \frac{V^2}{2g}$$

We may plot the total energy as shown in Figure k, where the variables are as follows:

- Z: Height of riverbed above a datum

- h: Depth of water

- V: Average velocity at a section

- $\dfrac{V^2}{2g}$: Kinetic energy head

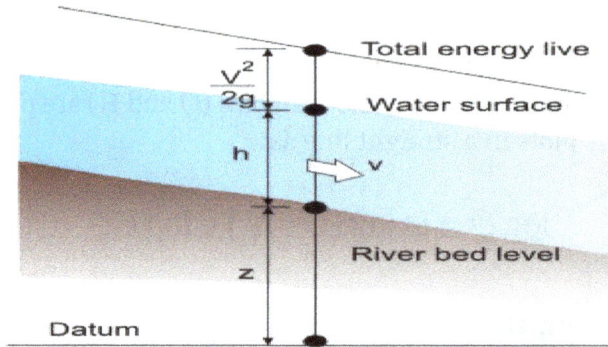

FIGURE k. Variation of total energy along a river

Since the cross section, bed slope and flow resistance vary along a river length, the depth and velocity would vary correspondingly. However, if a short stretch of a river section is taken, then the variations in riverbed, water surface and the total energy may be considered as linear (Figure l).

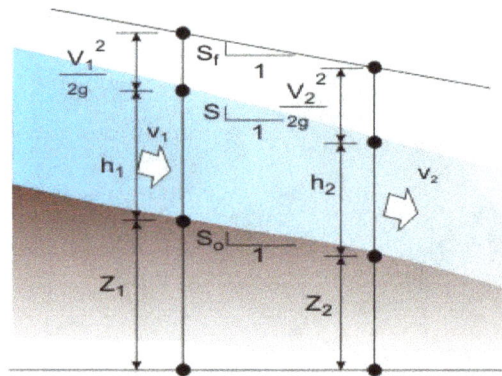

FIGURE l. Short section between two points along a river

In Figure l, three slopes have been marked, which are:

- S_o: Riverbed slope

- S: Water surface slope

- S_f: Energy surface slope

Since the total energy of flowing water reduces along the river length due to friction the "energy surface slope" is generally termed as the "friction slope". The energy loss in a river or an open channel occurs mostly due to the resistance at the channel sides, as the turbulent characteristics of the flowing water implies a smaller loss internally within the water body itself.

It has been nearly 200 years when scientists first attempted to mathematically express (or "model") the friction slope in terms of known variables like average velocity, cross section properties and riverbed slope. One of the earliest models for friction slope S_f or, in effect, the channel resistance was derived from the considerations of "uniform flow" (Figure m) where the flow variables and cross section are supposed to remain constant over a short reach.

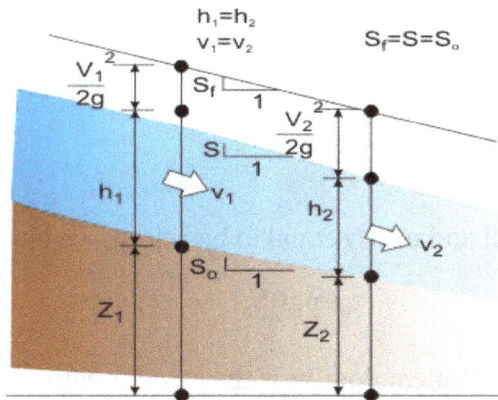

FIGURE m. Uniform flow along a channel reach

If we take small volume of fluid from these two sections we may make a free body diagram of the forces acting on it (Figure n).

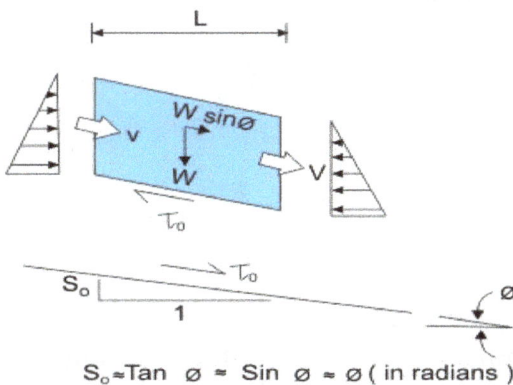

$$S_o \approx \text{Tan } \emptyset \approx \text{Sin } \emptyset \approx \emptyset \text{ (in radians)}$$

FIGURE n. Free body diagram of forces on a control volume in uniform flow

The variables represented in the figure are as follows

- W: Weight of water contained in the control volume

- V: Inflow velocity, which is the same as the outflow velocities

- θ : Angle of slope river bed, which is also equal to that water surface and friction slopes

- τ_0 : Shear stress due to friction acting on the control volume of fluid from the river bed and all along the periphery, though in Figure n only the resistance due to the riverbed is shown.

Equating the forces and noting that the inflowing and out flowing momenta are equal as well as the pressure forces at either end of the control volume one obtains:

$$\tau_0 PL = W \sin\theta = pgAL\sin\theta$$

Where the remaining variables are:

- P: wetted perimeter

- A: Cross section of flow area

- L: Length of control volume

Assuming θ to be very small and nearly equal to bed slope, we have

$$\tau_0 = pgRS_0$$

Assuming a state of rough turbulent flow, as is the case for natural rivers and channels, one may write

$$\tau_0 \alpha V^2 \qquad \text{or} \qquad \tau_0 = kV^2$$

Substituting into equation above,

$$V = \sqrt{\frac{\rho g}{k} RS_0}$$

This may be written as

$$V = C\sqrt{RS}$$

This is known as Chezy equation after the French hydraulic engineer. Antoine Chezy who first proposed the formula around 1768 while designing a canal for Paris water supply. The constant C in equation (8) actually varies depending on Reynolds number and boundary roughness.

In 1869, Swiss engineers, Ganguillet and Kutter proposed an elaborate formula for Chezy's C which they derived from actual discharge data from the river Mississippi and a wide range of natural and artificial channels in Europe. The formula, in metric units, is given as

$$C = 0.552 \left(\frac{41.6 + \dfrac{1.811}{n} + \dfrac{0.00281}{S_0}}{1 + \left[41.65 + \left(\dfrac{0.00281}{S_0} \right) \right] \dfrac{n}{\sqrt{R}}} \right)$$

Where n is a coefficient known as Kutter's n, and is dependent solely on the boundary roughness.

In 1889, Robert Manning's, an Irish engineer proposed another formula for the evaluation of the Chezy coefficient, which was later simplified to:

$$C = \frac{R^{1/6}}{n}$$

From Equation earlier, the Manning equation may be written as:

$$V = \frac{1}{n} R^{2/3} S_0^{1/2}$$

Where the Manning n is numerically equivalent to Kutter's n.

Uniform Flow in Channels of Simple Cross Section

For problems concerning the steady uniform flow in rivers and open channels, the Manning's equation is commonly used in India. The depth of water corresponding to a discharge in a channel or river under uniform flow conditions is called "normal depth". By combining the continuity equation with that of Mannings, one obtains

$$Q = \frac{1}{n} AR^{2/3} S^{1/2}$$

Where the variables have been defined in the earlier sections.

One may also write equation as follows

$$Q = K\sqrt{S}$$

Where $K = \dfrac{AR^{2/3}}{n}$ also called Conveyance, is often necessary to find out the normal depth of flow corresponding to a discharge Q, flowing in a channel for which equation may be rearranged as

$$AR^{2/3} = \frac{nQ}{S^{1/2}}$$

In equation (14), the right hand side terms are known where as those in left hand are unknown and are functions of water depth. For a few commonly encountered sections the parameters A and R are given in the table below.

	Rectangle	Trapezoid	Circle
Flow Area, A	$b.h$	$(b+my).y$	$\dfrac{1}{8}(\phi - sin\phi)$
Wetted Perimeter, P	$b+2h$	$b+2h\sqrt{1+m^2}$	$\dfrac{1}{2}\phi D$
Hydraulic Radius, R	$\dfrac{bh}{b+2h}$	$\dfrac{(b+my).y}{b+2h\sqrt{1+m^2}}$	$\dfrac{1}{4}\left(1-\dfrac{sin\,\phi}{\phi}\right)D$
Free surface width, B	b	$b+2mh$	$\left(sin\dfrac{\phi}{2}\right)D$

In the table, m stands for the side slope of a trapezoidal channel and < stands for the angle subtended at the centre by the water surface chord line.

As seen from the above table except for the very simple rectangular section it is not possible directly to evaluate h, corresponding to Q as the left hand side of equation earlier is nonlinear in terms of h. One way of solving is by Newton's method, where equation above is written as

$$f(h) = AR^{2/3} - \frac{nQ}{S^{1/2}} = 0$$

For using Newton's method the derivative of the function is required

$$f'(h) = \frac{d}{dh}\left(A\frac{A^{2/3}}{P^{2/3}} - \frac{nQ}{S^{1/2}}\right) = 0$$

$$f'(h) = \frac{5}{3}P^{-2/3}A^{2/3}\frac{dA}{dh} - \frac{2}{3}P^{-5/3}A^{5/3}\frac{dP}{dh}$$

$$f'(h) = \frac{5}{3}BR^{2/3} - \frac{2}{3}R^{2/3}\frac{dP}{dh}$$

Where we have used $\dfrac{dA}{dh} = B$. Similarly the expression $\dfrac{dP}{dh}$ may be evaluated for any section.

Starting with a realistic value h_i the iteration may be carried out as given below:

$$h^{i+1} = h^i - \frac{f(h^i)}{f'(h^i)}$$

Where h^{i+1} is the value of h at next iteration, which is an improvement of initial guess h^i. The iteration may be continued till a desired accuracy is achieved.

Uniform Flow in Channels of Compound Cross Section

A compound section may be defined as a section in which various portions of the cross-section have different flow properties, like surface roughness or channel depth. (Figure o).

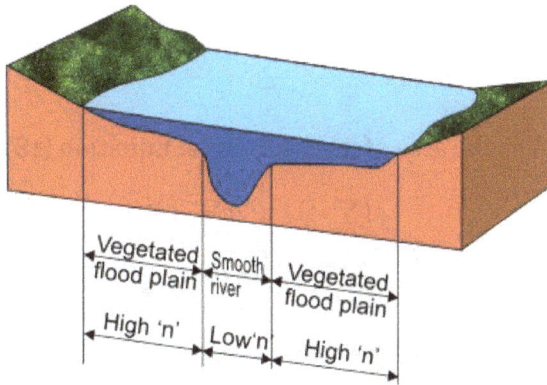

FIGURE o. A river in flood example of a compound flow section

In order to use the uniform flow formula in compound channels one way may be to divide the flow section into sub areas (Figure p) and treat the flow in each area separately.

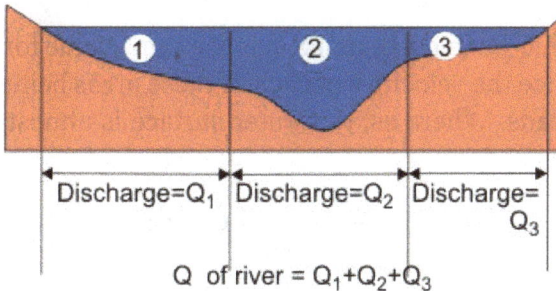

FIGURE p. Simple flow analysis method for compound channels

However, it has been found that this method may lead to errors by as much as ±20% or even more (Chadwick et al 2004). The error is largely due to the neglecting of mass and momentum interchange between adjacent sub-areas. The current solution would however be more complex by using a two or even three-dimensional model.

In another method, the energy coefficient (α) and friction slope S_f are evaluated in terms of conveyance K of the sub areas. With these expressions, the flow in compound section may be computed without knowing the individual flows in each sub area. For a compound channel divided into N sections. The energy coefficient, α, is found out as:

$$\alpha = \frac{\sum_{i=1}^{N} V_i^3 A_i}{V_m^3 \sum_{i=1}^{N} A_i}$$

Where V_m is the mean flow velocity in the entire section and is given as follows

$$\frac{\sum V_i A_i}{\sum}$$

Where $V_i = Q_i / A_i$ and A_i is the area of its ith sub-area. Equation (18) now can be written as

$$\alpha = \frac{\left(\sum Q_i^3 / A_i^2\right)\left(\sum A_i\right)^2}{\left(\sum Q_i\right)^3}$$

Now, the flow in sub-areas i may be written as

$$Q_i = K_i S_{fi}^{1/2}$$

$$S_{fi}^{1/2} = \frac{Q_i}{K_i}$$

Here, an assumption has been made that S_f has the same value for all sub-areas, which is not quite correct since the velocities of each of these areas being different, would not give equal velocity heads. Where as, the water surface is almost level over the entire cross section.

$$\frac{Q_1}{K_1} = \frac{Q_2}{K_2} = ----- = \frac{Q_n}{K_n} = \text{Constant} = S_f^{1/2}$$

It follows from equation (23) that

$$Q_1 = K_1 \frac{Q_n}{K_n}$$

$$Q_2 = K_2 \frac{Q_n}{K_n}$$

$$-$$

$$-$$

$$-$$

$$Q_n = K_n \frac{Q_n}{K_n}$$

Adding all the above equation yields

$$Q = \sum_{i=1}^{n} Q_i = \frac{Q_n}{K_n} \sum_{i=1}^{n} K_i$$

By substituting this expression for $Q_i = K_i \left(\dfrac{Q_n}{K_n} \right)$ into equation above and simplifying the equation, one obtains

$$\alpha = \frac{\left(\sum_{i=1}^{n} A_i \right)^2}{\left(\sum_{i=1}^{n} K_i \right)^3} , \sum_{i=1}^{n} \left(\frac{K_i^3}{A_i^2} \right)$$

Elimination of $\dfrac{Q_n}{K_n}$ from equations above and squaring both sides give

$$S_f = \left(\frac{\sum Q_i}{\sum K_i} \right)^2$$

$$S_f = \frac{Q^2}{\sum K_i^2}$$

Thus, expressions for α and S_f have been evaluated for any given stage without explicitly determining the flow in each sub areas, Q_i. In addition, equation (30) may be used in the procedure for determining varied flow profiles.

Non Uniform in Channels

There are quite a few examples of non-uniform flow in rivers or open channels that may be encountered by a water resources engineer. Some of these have been illustrated in Figure q.

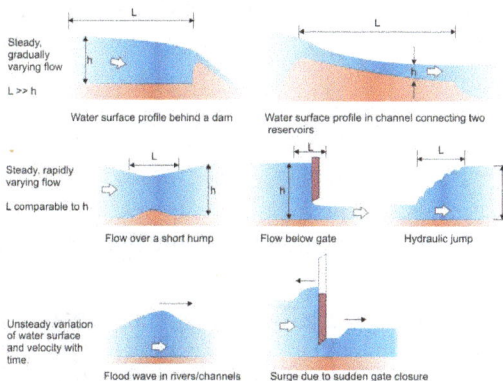

FIGURE q. Some example of flow situation encountered in practice.

In this lesson we shall discuss the procedure to evaluate water surface profiles for steady, gradually varying flow situations.

Non-uniform Gradually Varied Flow Calculation

A representative non-uniform gradually varied flow is shown in Figure r.

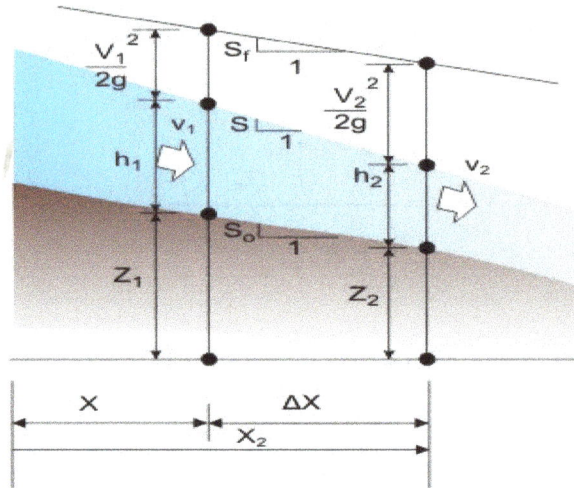

FIGURE r. Non-uniform gradually varied flow

Over the incremental distance Δx, the depth and velocity are known to change slowly. The slope of the energy grade line is designated as α in contrast to uniform flow, the slopes of the energy grade line, water surface, and channel bottom are no longer parallel. Since the changes in the water depth h and velocity V are gradual, the energy lost over the incremental Δx can be represented by manning equation. This means that equation, which is valid for uniform flow can also be used to evaluate S for a gradual varied flow situation, and that the roughness coefficients discussed.

Additional assumption includes a regular cross section, small channel slope, hydrostatic pressure distribution and one-dimensional flow.

Applying the equivalence of energy between locations 1 and 2, and assuming the loss term as h$_L$ given by S_f. Δx one obtains

$$Z_1 + h_1 + \alpha \frac{V_1^2}{2g} = Z_2 + h_2 + \alpha \frac{V_2^2}{2g} + S_f \Delta x$$

In the above equation, Δx is the distance between two consecutive sections x_1 and x_2 such that $\Delta x = x_2 - x_1$. The energy coefficient α has been used along with the $\frac{V^2}{2g}$ term, as it may be much different from 1.0 for natural sections. The term S$_f$ in equation above may be evaluated by the expression for uniform flow, equation above, where S$_o$ may be replaced by S$_f$. Since equation above relates the energy between the sections, S$_f$ may be taken either of the following:

$$\text{Arithmetic mean: } \overline{S_f} = \frac{1}{2}\left(S_{f1} + S_{f2}\right)$$

$$\text{Geometric mean: } \overline{S_f} = \sqrt{S_{f1}S_{f2}}$$

$$\text{Harmonic mean: } \overline{S_f} = \frac{2S_{f1}S_{f2}}{S_{f1}S_{f2}}$$

Where S_{f1} and S_{f2} are the friction slopes evaluated at section 1 and 2 by using the Mannings formula equation earlier.

Equation earlier may be used by starting from one end of the channel where the flow depth and velocity are known and working backward or forward in steps. Here, two, methods are used of which we shall discuss one, called the standard step method. Every popular computer program called HEC-2 developed by hydrologic engineering center of the US Army Corps of Engineers is based on this method.

In the standard step method, for any given discharge the depth of flow would be known at the control section. It is then required to calculate the depth of flow at the section immediately next to the control section. Two examples are illustrated in Figure s.

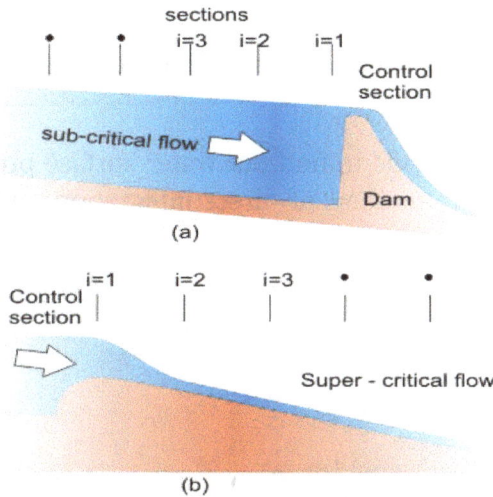

FIGURE s. Computation grid by the standard step method.

(a) profile behind a dam: calculation proceeds upstream from control section

(b) profile in a steep channel : calculation proceeds downstream from control section

The distance between the two successive sections (i and i+1) is taken as constant, say Δx. It may be observed from the Figure 19a since the water is flowing above the dam the water depth above the dam crest can be found out for the given discharge. Hence the water level at the control section just upstream of the dam is known. Similarly, in Figure 19b, since the water is flowing down from the reservoir into the steep channel critical depth corresponding to the given discharge would exist at the control section Here two, the water level at the control section is then known.

Starting at the control section (i =1), the total energy of water is found out to be

$$H_1 = Z_1 + h_1 + \alpha \frac{V_1^2}{2g}$$

Next, consider the first reach, that is, between sections i =1 and i =2. A depth of flow is assumed at section 2 and the energy there, that is,

$$H_2 = Z_2 + h_2 + \alpha \frac{V_2^2}{2g}$$

is evaluated. Now, one of the equations for finding $\overline{S_f}$ (the average friction slope) in the reach is found out by, say equation earlier.

As may be observed from Figure r the numerical value of H_2 found from equation above should be equal to that of h_1 found from equation $+S_f$. If the depth at the section 2 has been correctly assumed if the two don't match, a new depth h_2 is assumed and the calculations are repeated till the two values match.

Once a correct depth is found at section 2, a similar procedure is used to find the depth at section 3, and so on.

These are the two other methods to find out water surface profiles of gradually varied flow situations, namely; method of direct integration and method of graphical integration.

Gradually Varied Flow Profiles

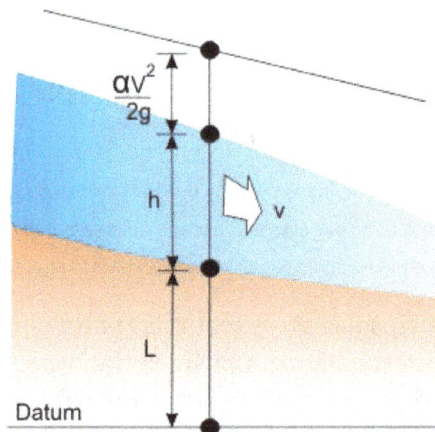

FIGURE t. Definition sketch

In many flow problems it is enough to make a qualitative sketch of water surface profile for a given flow that is taking place between two locations. It is not necessary therefore to find out the exact level of water at different points but the general shape of the free surface has to be drawn as accurately as possible. An analysis of water surface profile may be done by studying the governing equation, which can be derived from the sketch in Figure t.

The total energy H at a channel section is given as

$$H = Z + h + \alpha \frac{V^2}{2g}$$

Where

- H: Elevation of energy line above the datum

- Z: Elevation of channel bottom above datum

- h: Flow depth

- V: Mean flow velocity

- α : Velocity head coefficient

Considering x as the space coordinate, taken positive in the direction of flow one obtains by differentiating both sides of the equation above with respect to x and expressing V in terms of discharge Q.

$$\frac{dH}{dx} = \frac{dZ}{dx} + \frac{dh}{dx} + \alpha \frac{Q^2}{2g} \frac{d}{dx}\left(\frac{1}{A^2}\right)$$

Again, we know by definition:

$$\frac{dH}{dx} = -S_f$$

$$\text{And } \frac{dZ}{dx} = -S_o$$

In which

- S_f: Slope of the energy grade line

- S_o: Slope of the channel bottom.

The negative sign of S_f and S_o indicates that both H and Z decrease as x increases. In equation above an expression for the derivative of A^{-2} may be found out as follows:

$$\frac{d}{dx}\left(\frac{1}{A^2}\right) = \frac{d}{dA}\left(\frac{1}{A^2}\right)\frac{dA}{dx}$$

$$= \frac{d}{dA}\left(\frac{1}{A^2}\right)\frac{dA}{dh}\frac{dh}{dx}$$

$$= -\frac{2B}{A^3}\frac{dh}{dx}$$

Since $\dfrac{dA}{dh} = B$

By substituting equations earlier, and rearranging the resulting equation one obtains

$$\frac{dh}{dx} = \frac{S_o - S_f}{1-\left(\alpha BQ^2\right)/gA^3}$$

If the channel is not prismatic, then the cross sectional area A changes with distance, and may be expressed as:

$$\frac{dA}{dx} = \frac{\partial A}{\partial x} + \frac{\partial A}{\partial y}\frac{dh}{dx}$$

The above change would modify equations earlier accordingly.

We may express equation, which describes the variation of h with x, in terms of the Froude Number (Fr) if we note the following:

$$\frac{\alpha BQ^2}{gA^3} = \frac{(Q/A)^2}{(gA)/(\alpha B)} = Fr^2$$

Hence, equation may be written as

$$\frac{dh}{dx} = \frac{S_o - S_f}{1-Fr^2}$$

Equation above can give a general idea about the nature of the curve if one knows the relative inclinations of the channel bed slope and friction slope (S_f) and the Froude Number (Fr). This may be done by observing the water flow depth (h) with respect to normal depth (h_n) and critical depth (h_c) for a given discharge, the following figures show the relative changes of h_n and h_c as channel bed slope is increased gradually from horizontal. It may be observed that the for a given discharge h_c does not change but h_n goes on decreasing starting from an infinite value for a flat slope.

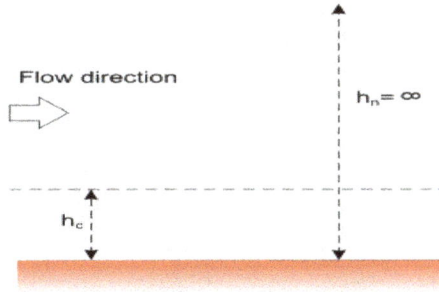

FIGURE u. Normal depth (h_n) and critical depth (h_c) on horizontal sloped bed

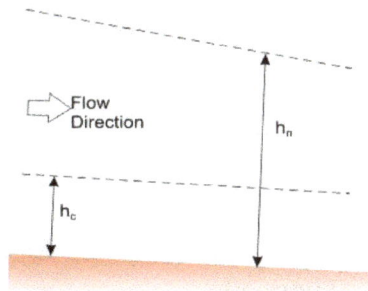

FIGURE 22. Slope increased slightly
$h_n > h_c$; mild slope

FIGURE v. Slope increased slightly $h_n > h_c$; mild slope

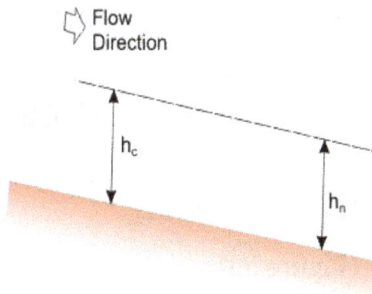

FIGURE w. Slope increased further $h_n > h_c$; critical slope

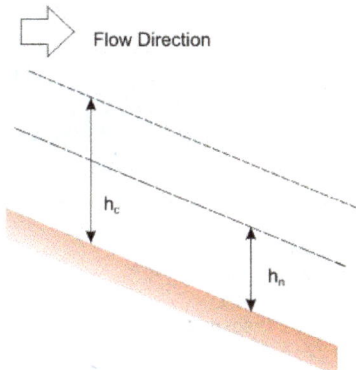

FIGURE x. Slope increased substantially $h_n > h_c$; Steep slope

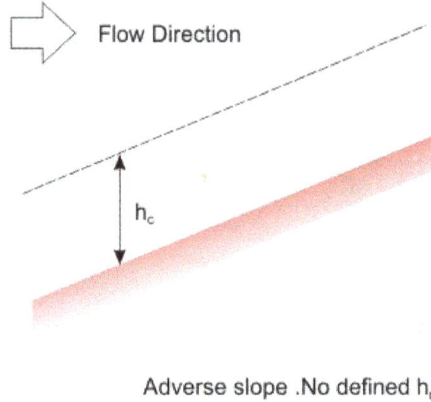

Adverse slope .No defined h_n

FIGURE y. Adverse slope. No defined h_n

In water resources projects, one generally encounters slopes of channels that are either of the following:

- Mild, where $h_n > h_c$ (Figure v)

- Steep, where $h_n < h_c$ (Figure x)

- Critical, where $h_n = h_c$ (Figure w)

- Flat, where $h_n = \infty$ (Figure u)

- Adverse, where the slope is reversed (Figure y)

For each of these slopes, the actual water surface would vary depending upon a control that exist either at the upstream or downstream end of the channel. Some examples of controls are given below

- Weir or spillway   (Figures z and a1)

- Gate   (Figure a2)

- Free overfall   (Figure a3)

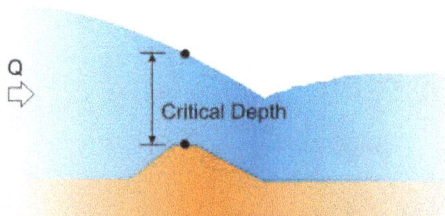

Weir length :L
Weir coefficient : C_d
V = Q/A
Q = Discharge
A = Area

FIGURE z. Flow over a weir

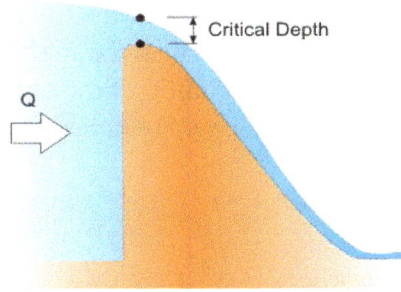

Spillway Length : L
V = Q/A
Q = Discharge
A = Area

FIGURE a1. Flow over a dam spilway

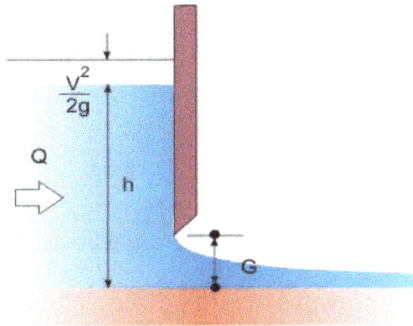

FIGURE a2. Flow below a gate. The depth of water just upstream of the gate (h) may be determined from the formula $Q=C.L.G\sqrt{H}$ where $H=h+v^2/2g$. The formula is valid for small gate openings

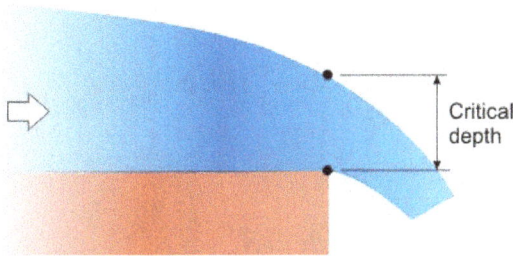

FIGURE a3. Free over fall

Apart from the above a normal depth may be assumed to exist within a very long channel, for which the conditions at the far end may be neglected (Figure a4).

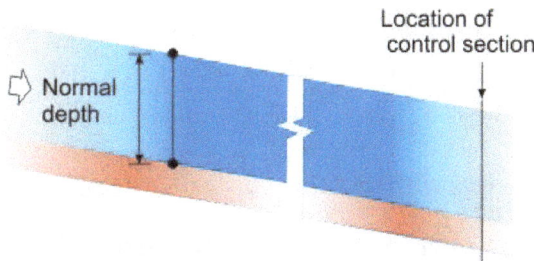

FIGURE a4. Flow in a long channel

The situation shown in Figure a4 is used often while analyzing flow in, say, at the tail end of long irrigation channels or in a long river. Examples illustrating the use of equation and a known control section in determining flow profiles where for a mildly slope channel. Similar profiles may be qualitatively sketched for other channels too.

Downstream Control Raising the Water Level Above Normal Depth

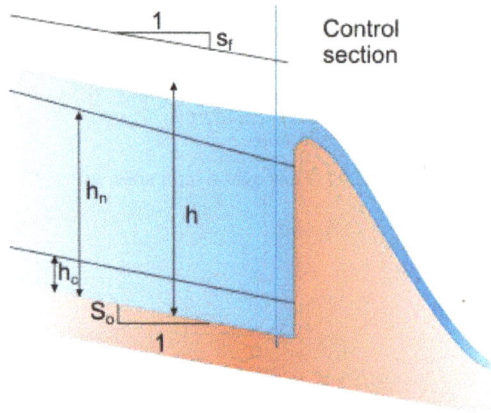

FIGURE a5. M1 type of water surface profile behind dam spillway

This situation is common for spillways of large dams. The flow profile in a mildly sloped channels where $h > h_n > h_c$ as shown in Figure 31 is known as the M_1 curve. Now, for uniform flow, $S_f = S = S_o$ when $h = h_n$. Hence it is clear from Mannings formula (equation 11), that for a given discharge, Q,

$$S_o > S_f \text{ if } h > h_n$$

Thus, in equation,

$$\frac{dh}{dx} = \frac{S_o - S_f}{1 - Fr^2}$$

The numerator is positive Fr<1 since h>h_c. Therefore, the denominator of equation above is positive as well. Hence, it follows from this eqn that

$$\frac{dh}{dx} = \frac{S_o - S_f}{1 - Fr^2} = \frac{+}{+} = +$$

This means that h increases with distance x.

Comparing with Figure a3 it may be inferred that quite some distance upstream of the spillway the flow depth nearly equals normal depth. And, since dh/dx for this profile is positive which means that the water depth goes on increasing towards the spillway, the flow depth becomes nearly horizontal. However very close to spillway the flow profile again changes which is due to the fact that the flow here is not really one-dimensional (Figure a6).

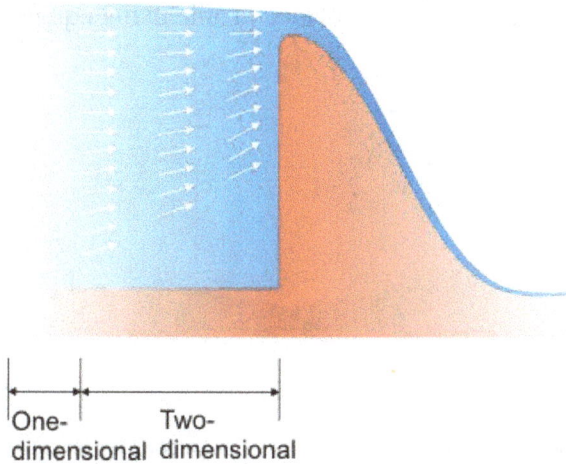

One-   Two-
dimensional dimensional

FIGURE a6. Flow situation very close to a spillway

Downstream Controlled Raising Water Level Above Critical Depth but Below Normal Depth

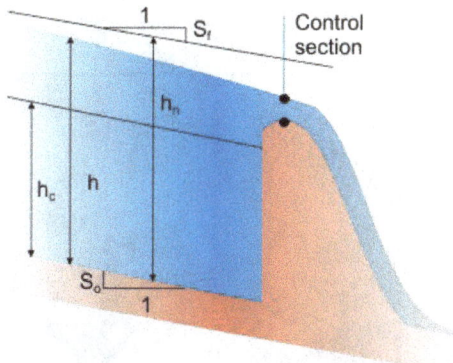

FIGURE a7. M2 type of water surface profile behind a short-heigt spillway

The flow profile in a mildly sloping channel where $h_n > h > h_c$, has been shown in Figure a7 is known as the M_2 curve. In this case $S_f > S_o$ since $h < h_n$ (from Mannings formula). Thus the numerator in equation above is negative. However, the denominator is positive, since Fr<1 because $h > h_c$ hence it follows from equation above that

$$\frac{dh}{dx} = \frac{S_o - S_f}{1 - Fr^2} = \frac{-}{-} = -$$

Thus h decreases as x increases for upstream of this spillway control section the flow depth would be asymptotic to normal depth h_n.

Upstream Control Causing Water Depth to be Less than Both Normal and critical Depths

This situation is shown in Figure a8 for flow taking place below a sluice gate. The reader is advised to check the trend of water surface profile using equation above in this case.

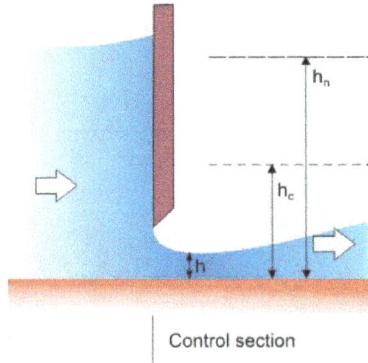

FIGURE a8. M3 type of water surface profile downstream of a gate

Wastewater

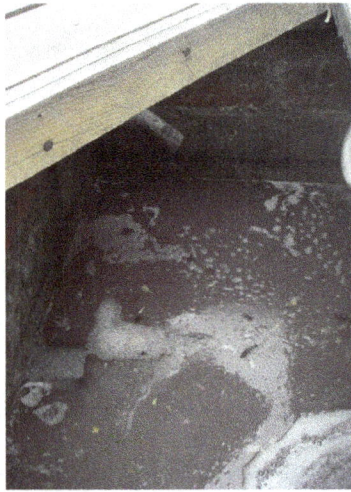

Greywater (a type of wastewater) in a settling tank

Wastewater, also written as waste water, is any water that has been adversely affected in quality by anthropogenic influence. Wastewater can originate from a combination of domestic, industrial, commercial or agricultural activities, surface runoff or stormwater, and from sewer inflow or infiltration.

Municipal wastewater (also called sewage) is usually conveyed in a combined sewer or sanitary sewer, and treated at a wastewater treatment plant. Treated wastewater is discharged into receiving water via an effluent pipe. Wastewaters generated in areas without access to centralized sewer systems rely on on-site wastewater systems. These typically comprise a septic tank, drain field, and optionally an on-site treatment unit. The management of wastewater belongs to the overarching term sanitation, just like the management of human excreta, solid waste and stormwater (drainage).

Sewage is a type of wastewater that comprises domestic wastewater and is therefore contaminated with feces or urine from people's toilets, but the term sewage is also used to mean any type of wastewater. Sewerage is the physical infrastructure, including pipes, pumps, screens, channels etc. used to convey sewage from its origin to the point of eventual treatment or disposal.

Sources

Wastewater can come from:

- Human excreta (feces and urine) often mixed with used toilet paper or wipes; this is known as blackwater if it is collected with flush toilets

- Washing water (personal, clothes, floors, dishes, cars, etc.), also known as greywater or sullage

- Surplus manufactured liquids from domestic sources (drinks, cooking oil, pesticides, lubricating oil, paint, cleaning liquids, etc.)

- Urban rainfall runoff from roads, carparks, roofs, sidewalks/pavements (contains oils, animal feces, litter, gasoline/petrol, diesel or rubber residues from tires, soapscum, metals from vehicle exhausts, etc.)

- Highway drainage (oil, de-icing agents, rubber residues, particularly from tires)

- Storm drains (may include trash)

- Manmade liquids (illegal disposal of pesticides, used oils, etc.)

- Industrial waste

- Industrial site drainage (silt, sand, alkali, oil, chemical residues);

    o   Industrial cooling waters (biocides, heat, slimes, silt)

    o   Industrial process waters

    o   Organic or biodegradable waste, including waste from abattoirs, creameries, and ice cream manufacture

    o   Organic or non bio-degradable/difficult-to-treat waste (pharmaceutical or pesticide manufacturing) o   Extreme pH waste (from acid/alkali manufacturing, metal plating)

o   Toxic waste (metal plating, cyanide production, pesticide manufacturing, etc.)

o   Solids and emulsions (paper manufacturing, foodstuffs, lubricating and hydraulic oil manufacturing, etc.)

o   Agricultural drainage, direct and diffuse

o   Hydraulic fracturing

o   Produced water from oil & natural gas production

Wastewater can be diluted or mixed with other types of water in the form of:

- Seawater ingress (high volumes of salt and microbes)

- Direct ingress of river water

- Rainfall collected on roofs, yards, hard-standings, etc. (generally clean with traces of oils and fuel)

- Groundwater infiltrated into sewage

After it has undergone some treatment, the "treated wastewater" remains, e.g.:

- Septic tank discharge

- Sewage treatment plant discharge

Constituents

The composition of wastewater varies widely. This is a partial list of what it may contain:

- Water (more than 95 percent), which is often added during flushing to carry waste down a drain;

- Pathogens such as bacteria, viruses, prions and parasitic worms;

- Non-pathogenic bacteria;

- Organic particles such as feces, hairs, food, vomit, paper fibers, plant material, humus, etc.;

- Soluble organic material such as urea, fruit sugars, soluble proteins, drugs, pharmaceuticals, etc.;

- Inorganic particles such as sand, grit, metal particles, ceramics, etc.;

- Soluble inorganic material such as ammonia, road-salt, sea-salt, cyanide, hydrogen sulfide, thiocyanates, thiosulfates, etc.;

- Animals such as protozoa, insects, arthropods, small fish, etc.;

- Macro-solids such as sanitary napkins, nappies/diapers, condoms, needles, children's toys, dead animals or plants, etc.;

- Gases such as hydrogen sulfide, carbon dioxide, methane, etc.;

- Emulsions such as paints, adhesives, mayonnaise, hair colorants, emulsified oils, etc.;

- Toxins such as pesticides, poisons, herbicides, etc.

- Pharmaceuticals and hormones and other hazardous substances

Quality Indicators

Any oxidizable material present in an aerobic natural waterway or in an industrial wastewater will be oxidized both by biochemical (bacterial) or chemical processes. The result is that the oxygen content of the water will be decreased. Basically, the reaction for biochemical oxidation may be written as:

Oxidizable material + bacteria + nutrient + $O_2 \rightarrow CO_2 + H_2O$ + oxidized inorganics such as NO_3^- or SO_4^{2-}

Oxygen consumption by reducing chemicals such as sulfides and nitrites is typified as follows:

$$S^{2-} + 2\,O_2 \rightarrow SO_4^{2-}$$

$$NO_2^- + \tfrac{1}{2}\,O_2 \rightarrow NO_3^-$$

Since all natural waterways contain bacteria and nutrients, almost any waste compounds introduced into such waterways will initiate biochemical reactions (such as shown above). Those biochemical reactions create what is measured in the laboratory as the biochemical oxygen demand (BOD). Such chemicals are also liable to be broken down using strong oxidizing agents and these chemical reactions create what is measured in the laboratory as the chemical oxygen demand (COD). Both the BOD and COD tests are a measure of the relative oxygen-depletion effect of a waste contaminant. Both have been widely adopted as a measure of pollution effect. The BOD test measures the oxygen demand of biodegradable pollutants whereas the COD test measures the oxygen demand of oxidizable pollutants.

The so-called 5-day BOD measures the amount of oxygen consumed by biochemical oxidation of waste contaminants in a 5-day period. The total amount of oxygen consumed when the biochemical reaction is allowed to proceed to completion is called the Ultimate BOD. Because the Ultimate BOD is so time consuming, the 5-day BOD has been almost universally adopted as a measure of relative pollution effect.

There are also many different COD tests of which the 4-hour COD is probably the most common.

There is no generalized correlation between the 5-day BOD and the ultimate BOD. Similarly there is no generalized correlation between BOD and COD. It is possible to develop such correlations for specific waste contaminants in a specific wastewater stream but such correlations cannot be generalized for use with any other waste contaminants or wastewater streams. This is because the composition of any wastewater stream is different. As an example an effluent consisting of a solution of simple sugars that might discharge from a confectionery factory is likely to have organic components that degrade very quickly. In such a case, the 5 day BOD and the ultimate BOD would be very similar since there would be very little organic material left after 5 days. However a final effluent of a sewage treatment works serving a large industrialised area might have a discharge where the ultimate BOD was much greater than the 5 day BOD because much of the easily degraded material would have been removed in the sewage treatment process and many industrial processes discharge difficult to degrade organic molecules.

The laboratory test procedures for the determining the above oxygen demands are detailed in many standard texts. American versions include the "Standard Methods for the Examination of Water and Wastewater."

Treatment

There are numerous processes that can be used to clean up wastewaters depending on the type and extent of contamination. Wastewater can be treated in wastewater treatment plants which include physical, chemical and biological treatment processes. Municipal wastewater is treated in sewage treatment plants (which may also be referred to as wastewater treatment plants). Agricultural wastewater may be treated in agricultural wastewater treatment processes, whereas industrial wastewater is treated in industrial wastewater treatment processes.

For municipal wastewater the use of septic tanks and other On-Site Sewage Facilities (OSSF) is widespread in some rural areas, for example serving up to 20 percent of the homes in the U.S.

One type of aerobic treatment system is the activated sludge process, based on the maintenance and recirculation of a complex biomass composed of micro-organisms able to absorb and adsorb the organic matter carried in the wastewater. Anaerobic wastewater treatment processes (UASB, EGSB) are also widely applied in the treatment of industrial wastewaters and biological sludge. Some wastewater may be highly treated and reused as reclaimed water. Constructed wetlands are also being used.

Disposal

In some urban areas, municipal wastewater is carried separately in sanitary sewers and runoff from streets is carried in storm drains. Access to either of these is typically through a manhole. During high precipitation periods a combined sewer overflow can occur, forcing untreated sewage to flow back into the environment. This can pose a serious threat to public health and the surrounding environment.

Industrial wastewater effluent with neutralized pH from tailing runoff in Peru.

Sewage may drain directly into major watersheds with minimal or no treatment but this usually has serious impacts on the quality of an environment and on the health of people. Pathogens can cause a variety of illnesses. Some chemicals pose risks even at very low concentrations and can remain a threat for long periods of time because of bioaccumulation in animal or human tissue.

Wastewater may be pumped underground through an injection well.

Reuse

Treated wastewater can be reused in industry (for example in cooling towers), in artificial recharge of aquifers, in agriculture and in the rehabilitation of natural ecosystems (for example in Florida's Everglades). In rarer cases it is also used to augment drinking water supplies.

There are several technologies used to treat wastewater for reuse. A combination of these technologies can meet strict treatment standards and make sure that the processed water is hygienically safe, meaning free from bacteria and viruses. The following are some of the typical technologies: Ozonation, ultrafiltration, aerobic treatment (membrane bioreactor), forward osmosis, reverse osmosis, advanced oxidation.

Some water demanding activities do not require high grade water. In this case, wastewater can be reused with little or no treatment. One example of this scenario is in the domestic environment where toilets can be flushed using greywater from baths and showers with little or no treatment.

Gardening and Agriculture

There are benefits of using recycled water for irrigation, including the lower cost compared to some other sources and consistency of supply regardless of season, climatic conditions and associated water restrictions. Irrigation with recycled wastewater can also serve to fertilize plants if it contains nutrients, such as nitrogen, phosphorus and potassium.

Around 90% of wastewater produced globally remains untreated, causing widespread water pollution, especially in low-income countries. Increasingly, agriculture is using untreated wastewater for irrigation. Cities provide lucrative markets for fresh produce, so are attractive to farmers. However, because agriculture has to compete for increasingly scarce water resources with industry and municipal users, there is often no alternative for farmers but to use water polluted with urban waste directly to water their crops.

Health Risks

There can be significant health hazards related to using untreated wastewater in agriculture. Wastewater from cities can contain a mixture of chemical and biological pollutants. In low-income countries, there are often high levels of pathogens from excreta, while in emerging nations, where industrial development is outpacing environmental regulation, there are increasing risks from inorganic and organic chemicals. The World Health Organization, in collaboration with the Food and Agriculture Organization of the United Nations (FAO) and the United Nations Environmental Program (UNEP), has developed guidelines for safe use of wastewater in 2006. These guidelines advocate a 'multiple-barrier' approach to wastewater use, for example by encouraging farmers to adopt various risk-reducing behaviours. These include ceasing irrigation a few days before harvesting to allow pathogens to die off in the sunlight, applying water carefully so it does not contaminate leaves likely to be eaten raw, cleaning vegetables with disinfectant or allowing fecal sludge used in farming to dry before being used as a human manure.

Legislation

European Union

Council Directive 91/271/EEC on Urban Wastewater Treatment was adopted on 21 May 1991, amended by the Commission Directive 98/15/EC. Commission Decision 93/481/EEC defines the information that Member States should provide the Commission on the state of implementation of the Directive.

United States

The Clean Water Act is the primary federal law in the United States governing water pollution.

Philippines

In the Philippines, Republic Act 9275, otherwise known as the Philippine Clean Water Act of 2004, is the governing law on wastewater management. It states that it is the country's policy to protect, preserve and revive the quality of our fresh, brackish and marine waters, for which wastewater management plays a particular role.

References

- Di Luzio, Frank C. (January 1967). "Water Pollution Control: An American Must". Journal (Water Pollution Control Federation). Water Environment Federation. 39 (1): 1–7. JSTOR 25035710

- G. Allen Burton, Jr., Robert Pitt (2001). Stormwater Effects Handbook: A Toolbox for Watershed Managers, Scientists, and Engineers. New York: CRC/Lewis Publishers. ISBN 0-87371-924-7

- Wachman, Richard (2007-12-09). "Water becomes the new oil as world runs dry". The Guardian. London. Retrieved 2015-09-23

- Pfunt,H; Houben,G; Himmelsbach,T (2016). "Numerical modeling of fracking fluid migration through fault zones and fractures in the North German Basin". Hydrogeology Journal. 24 (6): 1343–1358. doi:10.1007/s10040-016-1418-7

- Goel, P.K. (2006). Water Pollution - Causes, Effects and Control. New Delhi: New Age International. p. 179. ISBN 978-81-224-1839-2

- "Report Offers First Worldwide Estimate of Investments in Combating Water Pollution". American Association for the Advancement of Science. Retrieved April 18, 2011

- Suthar, S; Bishnoi, P; Singh, S; et al. (2009). "Nitrate contamination in groundwater of some rural areas of Rajasthan, India". Journal of Hazardous Materials. 171 (1–3): 189–199. doi:10.1016/j.jhazmat.2009.05.111

- Kennish, Michael J. (1992). Ecology of Estuaries: Anthropogenic Effects. Marine Science Series. Boca Raton, FL: CRC Press. pp. 415–17. ISBN 978-0-8493-8041-9

- World Health Organization (WHO) (2006). "Protecting Groundwater for Health - Understanding the drinking-water catchment" (PDF). Retrieved 20 March 2017

- Singh, B; Singh, Y; Sekhon, GS (1995). "Fertilizer-N use efficiency and nitrate pollution of groundwater in developing countries". Journal of Contaminant Hydrology. 20 (3–4): 167–184. doi:10.1016/0169-7722(95)00067-4

- Laws, Edward A. (2000). Aquatic Pollution: An Introductory Text. New York: John Wiley and Sons. p. 430. ISBN 978-0-471-34875-7

- ATSDR (US Agency for Toxic Substance & Disease Registry) (2008). "Follow-up Health Consultation: Anniston Army Depot." (PDF). Retrieved 18 March 2017

- Smith, M; Cross, K; Paden, M; Laben, P, eds. (2016). Spring - managing groundwater sustainably (PDF). IUCN. ISBN 978-2-8317-1789-0

- "Texas fracking site that spilled 42,000 gallons of fluid into residential area hopes to reopen". RT International. Retrieved 2016-05-07

PERMISSIONS

All chapters in this book are published with permission under the Creative Commons Attribution Share Alike License or equivalent. Every chapter published in this book has been scrutinized by our experts. Their significance has been extensively debated. The topics covered herein carry significant information for a comprehensive understanding. They may even be implemented as practical applications or may be referred to as a beginning point for further studies.

We would like to thank the editorial team for lending their expertise to make the book truly unique. They have played a crucial role in the development of this book. Without their invaluable contributions this book wouldn't have been possible. They have made vital efforts to compile up to date information on the varied aspects of this subject to make this book a valuable addition to the collection of many professionals and students.

This book was conceptualized with the vision of imparting up-to-date and integrated information in this field. To ensure the same, a matchless editorial board was set up. Every individual on the board went through rigorous rounds of assessment to prove their worth. After which they invested a large part of their time researching and compiling the most relevant data for our readers.

The editorial board has been involved in producing this book since its inception. They have spent rigorous hours researching and exploring the diverse topics which have resulted in the successful publishing of this book. They have passed on their knowledge of decades through this book. To expedite this challenging task, the publisher supported the team at every step. A small team of assistant editors was also appointed to further simplify the editing procedure and attain best results for the readers.

Apart from the editorial board, the designing team has also invested a significant amount of their time in understanding the subject and creating the most relevant covers. They scrutinized every image to scout for the most suitable representation of the subject and create an appropriate cover for the book.

The publishing team has been an ardent support to the editorial, designing and production team. Their endless efforts to recruit the best for this project, has resulted in the accomplishment of this book. They are a veteran in the field of academics and their pool of knowledge is as vast as their experience in printing. Their expertise and guidance has proved useful at every step. Their uncompromising quality standards have made this book an exceptional effort. Their encouragement from time to time has been an inspiration for everyone.

The publisher and the editorial board hope that this book will prove to be a valuable piece of knowledge for students, practitioners and scholars across the globe.

Index